D1015769

ENERGY COSTS AND COSTING

A Selected, Annotated Bibliography

by

Emanuel Benjamin Ocran

The Scarecrow Press, Inc.
Metuchen, N.J., & London
1983

Library of Congress Cataloging in Publication Data

Ocran, Emanuel Benjamin.
Energy costs and costing.

Includes indexes.
1. Power resources--Costs--Bibliography. 2. Fuel--Prices--Bibliography. 3. Energy industries--Costs--Bibliography. 4. Energy industries--Prices--Bibliography. 5. Power resources--United States--Costs--Bibliography. 6. Fuel--Prices--United States--Bibliography. 7. Energy industries--United States--Costs--Bibliography. 8. Energy industries--Prices--United States--Bibliography. I. Title.
Z5853.P83033 1983 016.3384'3621042 83-8610
[HD9502.A2]

ISBN 0-8108-1631-8

Manufactured in the United States of America

CONTENTS

INTRODUCTION

Energy has always been an important commodity in everyday life. As an industry, its importance was made abundantly clear by the 1973 oil crisis which followed the Arab-Israeli war. Since then, each succeeding year has added to the toll of energy costs and these costs have caused rising prices in industrial production; hence the very high price of goods in the shops. The rise in the cost of transport services is also caused largely by the rise in energy costs. To control energy costs effectively, it is useful to have access to literature dealing with energy costs. This bibliography is therefore intended to provide such access. It lists books, reports and journal articles on energy costs and costing. The predominance of journal articles listed shows the dynamic nature of this source for current and useful information on energy. Energy disciplines in the bibliography are: Coal, Electricity, Energy (General), Fuels, Gas, Nuclear Energy, Oil, Solar Energy, and Wind Energy. Entries under these disciplines are grouped under the following subject headings, where applicable: Accounting, Costs, Finance, Insurance, Management, Pricing, Statistics, and Taxation.

Selection of Entries

Entries for the bibliography are very selective and are intended to be of use to all those concerned with the rising cost of energy. Selection is international in scope and the following are the sources consulted for selection:

Applied Science and Technology Index.
New York: H. W. Wilson, 1967+

British Technology Index.
London: Library Association, 1962+

Engineering Index.
New York: Engineering Index, Inc., 1928+

Direct scanning of current issues of journals not yet indexed.

Level of Use

The bibliography is designed primarily for the following users:

1. Librarians--as a tool for answering queries on energy costs and costing.
2. Consultants on energy--to discover parallels and leads as aids in consultancy.
3. Industry--to track down other means of achieving energy cost control.
4. Students of economics.
5. Management.
6. Researchers and specialists in the oil industry.
7. Transport managers and operators.
8. Those responsible for structuring corporate plans with special emphasis on energy costs and costing.

Arrangement

Under the chosen subject headings, entries are arranged alphabetically by title. These entries are numbered consecutively. In the case of journal articles, citations are in parentheses, consisting of the journal title, volume or issue number, date of issue and inclusive pagination; e.g., (In Elect. World, vol. 184, Nov. 15, 1975, pp. 44-47)

Annotations

Most of the entries are annotated. The very few which are not annotated were beyond the reach of the author for examination. Annotations cover the following:

1. Contents of books and articles
2. Charts and tables
3. A summary of the listed item.

The choice of content or paragraph headings of an article or paper to form the annotation is intended to present detailed information on what the article or paper is about. Content headings have provided a rich field for the subject index to the bibliography. The following is an example of an annotated entry:

Campus utility costs, by B.O. Turner. (In Heating & Piping, vol. 45, Nov. 1973, pp. 85-93)
Presents studies at three Utah institutions of higher learning. Yields comprehensive data that can be used to estimate utility consumptions and costs for existing buildings not metered, buildings under construction, and future buildings.... # Tables: 1, Heating--annual energy consumption, cost and peak demand comparisons; 2, Cooling--annual energy consumption and cost comparisons; 3, Domestic water--annual consumption, cost and peak demand comparisons; 4, Total electricity--annual consump-

tion, cost and peak demand comparisons; 5, Electricity for mechanical and for lighting--annual consumption and cost comparison.

Charts and tables giving invaluable information are treated as annotations. Sub-titles are treated as annotations only when they are judged to be helpful in clarifying the main title. The symbol # is used as a mark of emphasis for a complete ordered field in the annotation.

Name Index

The name index is a source of reference for known authors listed in the bibliography. Initials of authors are given instead of the full forename. Included in the name index are corporate authors.

Subject Index

This is a detailed index structured on the chain indexing procedure. Most index entries are made up of three subject concepts, which, by the use of colons, form a chain of three links. Index concepts for leading links in the chain are derived mainly from titles and annotations of listed items. The following entry under Costs in Fuels illustrates the structuring of subject entries:

486 Fuel costs hinder Olympic's return to profitability, by H. Lefer. (In Air Transport World, Nov. 1979, pp. 78-80, 83-84, 86)
Describes the optimism of management that new streamlined organization, improved labor relations and a billion dollar re-equipment plan will pay off.

For this item the following entries appear in the subject index:

Labor relations: Costs: Fuels 486
Management: Costs: Fuels 486
Olympic airline: Costs: Fuels 486
Profits: Costs: Fuels 486

The subject to be sought is the leading concept or link in the chain. The second or middle link is the broad subject heading which, in the examples given, is Costs, to which the leading concept is related. In cases where the sought concept is combined with the subject heading to form a unified concept, the middle link in the chain is dropped and the sought concept linked directly to the third link, which is the energy discipline. For example:

Life-cycle costs: Fuels 498
Lighting cost analysis: Electricity 80

Load management: Electricity 171
Load management: Fuels
Low costs: Electricity 100
Marginal cost pricing: Energy 443
Marginal costs: Electricity 101

Cross-references in the subject index provide additional information which otherwise might escape attention.

Appendixes

Appendix A: Statistical References consists mainly of reference tools and other publications containing useful statistical information on energy.

Appendix B lists titles of journals cited in the bibliography. Where an abbreviated form of the title has been cited, the full title is given in the appendix.

Acknowledgments

I am most grateful to the following libraries for making it possible to compile this bibliography: The National Reference Library, Holborn; The Civil Aviation Authority Central Library; and the Department of Energy Library. I am also grateful to my family for giving me moral support at every stage of my work.

E. B. OCRAN

I. COAL

Costs

1 Analysis of effects of legislation upon reserves and profits in contour surface mining, by J. M. Mutmansky. (In Coal Age, vol. 79, Sept. 1974, pp. 104-108)
Contents: Changes in reserves and profits; Changes in overburden-dependent costs; Changes in coal-dependent costs. # Tables: Numerical summary of results of changes in reclamation costs; Numerical summary of results of changes in overburden-dependent costs; Numerical summary of results of changes in coal-dependent costs.

2 Bank takes an optimistic view of the coal industry, by P. R. Lesutis. (In Coal Age, vol. 77, Sept. 1972, pp. 94-97)
Tables: 1, Estimated world coal reserves; 2, Production in major foreign countries; 3, Relative strength of US productivity; 4, Coal as a percentage of fuels used in steam generating plant. # Most of the coal produced in the US is sold under long-term contract with price escalation provision to cover cost increases during the life of the contract.

3 Capital and operating cost for new properties, by N. Robinson. (In Min. Cong. J., vol. 55, Sept. 1969, pp. 72-75)
Presents an analysis of operating costs based on original capital investment. # Tables: Estimated capital costs; Estimated cost of production; Estimated labor budget.

4 Changing parts ahead of time saves production costs in the long run. (In Coal Age, vol. 80, Apr. 1975, pp. 138)
Discusses the cost-saving benefits of preventive maintenance.

5 Coal Age mining guide book: coal's challenge. (In Coal Age, vol. 74, July 1970, pp. 126-127)
Discusses investment and the control of costs.

6 Coal Age mining guide book: continuing challenge. (In Coal Age, vol. 74, July 1969, pp. 144-145)
Lists as major goals for investing for profit as: Lower costs, Lower labor costs, Less downtime and lower maintenance cost. # Under controlling cost, it discusses the

following points: Effective control, Budgeting, Data processing, Methods revision, and Operations research.

7 Coal conversion plants cost five times refineries, by W. L. Nelson. (In Oil & Gas J., vol. 77[16], Apr. 23, 1979, p. 91)
Question: What will be the cost of coal conversion plants to replace oil and gas fuels?

8 Cost and effect of labour and equipment in the modern day coal mine, by A.R. MacLean. (In Can. Min. & Met. Bull., vol. 62, Oct. 1969, pp. 1101-1107)
Outlines the changes in modern-day coal mining that have been brought about by the increased cost of labor and the use of new and highly efficient equipment.

9 Cost estimation for coal prep financing, by G. H. K. Schenck. (In Coal Min. Process, vol. 15[8], Aug. 1978, pp. 64, 66-68, 70, 88, 90)
Outlines how to use cost indexes, empirical exponential formulae and simulation methods to derive realistic cost estimates for coal preparation plants.

10 Cost of solids-liquids separation methods in coal liquefaction, by K. Migut and S. Kasper. (In Am. Chem. Soc. Div. Fuel Chem. Papr., vol. 22[7], 1977, pp. 51-55)
Internal studies have shown the cost of such separation to vary from 5 to 20 percent of the total product oil cost. # Presents the result of an economic study in which three solids removal systems in typical liquefaction processes were examined in an attempt to find a reliable, cost-effective scheme.

11 Economic comparison of coal feeding systems in coal gasification: Lock Hopper vs. Slurry, by W.C. Morel. (In Am. Chem. Soc. Div. Fuel Chem. Papr., vol. 22[7], 1977, pp. 155-164)
Estimates are based on January 1977 cost indexes. Average selling prices of the gas were determined by using discounted cash flow rates of 12, 15, and 20 percent at various costs.

12 Economic considerations in the production and preparation of coal for the carbonization market, by G. Blackmore. (In Can. Min. Met. Bull., vol. 63, 1970, pp. 65-70)
Analyzes the labor costs of coal production over past years, and suggests that forward planning should provide for capital investment to make a traditionally labor-dense industry into a capital-dense industry.

13 Economic screening evaluation of upgrading coal liquids to turbine fuels. Paulsboro, NJ: Mobile Res. & Dev. Corp., 1978. 73p.

Costs were estimated for a 20MB/SD hydrotreater including waste water treater and sulphur plant using both utility and equity methods for financing.

14 Economic state of the coal industry, by H. S. Richey. (In Min. Cong. J., vol. 58, Feb. 1972, pp. 55-58)
"The coal energy indsutry--at its moment of greatest opportunity--is beset by problems from within and without. We have financial problems, safety problems, labor problems, and certainly production problems."

15 Economics of six coal-to-SNG processes, by R. Detman. (In Hydrocarbon Process, vol. 56[3], Mar. 1977, pp. 115-118)
Presents results of comparative economic analysis of six coal-to-SNG processes which set the range of gas costs and identify areas for immediate process development.

16 Estimate of reclamation costs resulting from federal law, by N. B. Pundari and J. A. Coates. (In Coal Age, vol. 80, Apr. 1975, pp. 127-131)
Discusses tangible and intangible costs. # Table: Mine model used for estimating tangible costs--shows estimates of tangible, intangible and total reclamation costs (per acre and per ton).

17 Heading off the high cost of cleaning up mine openings, by R. Reltew and E. E. Murphy. (In Coal Age, vol. 80, May 1975, pp. 70-72.
Defines the problem of sloughing, examines the present costs, and offers suggestions aimed at lowering overall cleanup expenditures.

18 Mining coal with scrapers ... by R. C. Gessel. (In World Coal, vol. 1[7], Sept. 1975, pp. 16-20)
Charts: Cost per bank cubic yard (in dollars); Cost per ton (in dollars).

19 More escalation seen for coal costs, by L. G. Hauser and R. F. Potter. (In Elec. World, vol. 174, Aug. 15, 1970, pp. 45-48.
An evaluation of future generation plants which includes a case analysis of expected fuel price trends. Also, a study of the effects of escalation on future utility fuel costs. # Partial Tables: Cost indices for coal industry, 1950-1965; Projection of fuel prices, 1967-1980; Fuel costs in ¢/10^6 Btu, with firm enrichment price.

20 Navajo mine converts to truck/rail, cutting haulage operating costs 25%, by D. Jackson. (In Coal Age, vol. 80, May 1975, pp. 52-53)
The rail portion, only 8½ miles long, ultimately reaches 50 miles and will handle most of the coal produced at the mine.

21 Organise your attack against high cost, by W. F. Saalbach. (In Coal Age, vol. 78, Aug. 1973, pp. 96)
Suggests that a manager can attack high cost and waste by searching for waste systematically in six general areas: Waste of time; Waste of materials and supplies; Waste of manpower and skills; Waste of utilities and services; Waste of machinery and equipment; and Waste of space.

22 Planning, financing and installing a new deep mine in the Beckley coal bed, by C. H. Williams. (In Min. Cong. J., vol. 60, Aug. 1974, pp. 42-47)
Partial Contents: Shaft slop cost put at $2.5 million; Costs figure out at $21.84 per annual ton.

23 Raise boring cuts shaft cost, by R. B. Hewes. (In Coal Age, vol. 75, July 1970, pp. 62-64)
Discusses how the new approach to the old problem of putting in a new air shaft has paid off in lower overall cost and a significant saving in time at Hanna Coal's Rose Valley mine.

24 Recent developments in conveyer vs. trunk haulage, by W. Laird. (In Min. Cong. J., vol. 60, Mar. 1974, pp. 37-43)
Examines the capital cost, some advantages and disadvantages of six systems for main line transportation of coal and rock....# Exhibits 1-6, Estimated transportation system cost; 7, Estimated cost of extending main entry haulage facilities to 25,000 ft.; 8, Equipment cost comparison....; 9, Estimated total transportation system costs for the main entry lengths of 5,000, 25,000, and 35,000 ft.

25 Transportation costs may delay utility coal use, by J. T. Miskell. (In Energy, vol. 5[1], 1980, pp. 10-11)
Escalating transportation costs from coal fields in the West to users in the South and Southwest are a continued problem.

26 Will underground coal gasification ever be commercialized in the USA? by R. L. Arscott and A. M. Garon. Washington, D. C.: Dept. of Energy, 1977. [15]p.
Analyzes the cost of underground process and the market options for utilization of the products.

27 Yardstick of productivity: Is it high tons per man-day, or is it low cost per ton? (In Coal Age, vol. 80, July 1975, pp. 93-98)
Contends that it is both. More tons per man-day can be achieved through application of new mining concepts. Low cost per ton means minimizing downtime and using horsepower instead of manpower whenever possible.

Finance

28 The annual audit: a way to sharpen dollar decisions. (In Coal Age, vol. 77, Nov. 1972, pp. 99-100)

Discusses the need to tuck more hard cash away--with a plan for yearly savings; and the need for more hard decisions to be made with a plan encompassing long-range needs.

29 Coal in the UK. London: Department of Energy, 1977. 18p. (Fact sheet, 4)
Partial Contents: National Coal Board--finance; Supply and demand--cost of production, pricing.# Describes the coal industry and the major contribution it makes in supplying the UK's overall energy needs.

30 Coal industry tripartite group sub-committee on the South Wales coal fields. Department of Energy, 1979. [30]p.
Partial Contents: Exploration and investment; Restructuring to achieve financial viability.

Management

31 Changed role of the foreman, by W.F. Saalbach. (In Coal Age, vol. 74, Sept. 1969, pp. 106-107)
Contents: The need for today's foreman; How higher management can help; Summary for the foreman; Summary for higher management.

32 Coal: still the top drawing card among new mining graduates, by T.M. Li. (In Coal Age, vol. 78, Apr. 1973, pp. 85-87)
Contents: Behind the boom in mining graduates; Coal tops list of opportunities; Government salaries improving; How are schools preparing students for mining; Trends in minerals engineering curricula.

33 Coal age mining guidebook: fundamental responsibilities for modern coal managers. (In Coal Age, vol. 76, July 1971, pp. 154-155)
Topics Discussed: A wise investment policy--lower costs, lower labor cost, machine utilization; A cost control program--budgeting, standard data.

34 Economics of in situ coal recovery, by C.S. Goddin, and others. (In Hydrocarbon Process, vol. 55, July 1976, pp. 109-111)
In situ coal gasification may be cost-competitive with other available conversion systems based on preliminary studies of the Lawrence Livermore Laboratory concept. The cost range for producing pipeline gas compares well with estimates for producing gas using the Lurgi process fed with strip-mined coal. Several critical parameters which significantly affect in situ costs must be defined by field tests. Includes a table of financial analysis.

35 Evaluation of coal for conversion to liquid hydrocarbons, by J.F. Cudmore. (In Austr. Inst. Min. & Metal. Symp. Ser. 18, Aug. 1977, pp. 146-158)

Outlines alternative processes for converting coal to liquid products and their stages of development. Factors involved in assessing investment and operating costs for modern hydrogenation-based coal refineries of different capacities are described.

36 Fighting the productivity slump with ACT, by R.A. Mason. (In Coal Min. Process., vol. 4[10], Oct. 1977, pp. 92, 94, 122)
Analyzes a production-orientated management system called accountability checkpoint technique--adopted in 18 coal producing companies and instituted in over 100 of their mines.

37 Give us the money and we'll give you the coal, by D. Booth. (In Engineer, vol. 235, Nov. 23, 1970, pp. 32-33, 35)
Coal Board Chief Derek Ezra in an interview vindicates his claim for more government money.

38 History, management philosophy: Eastern Associated Coal Corporation. (In Coal Age, vol. 74, Oct. 1969, pp. 101-107)
Discusses, among other topics, the capital expenditure of the corporation.

39 Innovations in the use of manpower in the coal industry, by B.S. Britton. (In Min. Congr. J., vol. 58, Nov. 1972, pp. 23-25)
Presents a formula for production and modifies it as innovations: 1, Miners are people; 2, Foremen are members of management; 3, Management by participation; 4, Executive development; 5, Innovate, create, be different.

40 Manager's requirement for maintenance, by W.G. Kegel. (In Min. Cong. J., vol. 57, June 1971, pp. 27-31)
Discusses coal mine maintenance as a key to successful operation.

41 Motivation is a perennial management problem, by W.F. Saalbach. (In Coal Age, vol. 79, Feb. 1974, pp. 93-94)
Sets out the goals for orientation. New employees should expect, and receive, superior orientation training.

42 New management team sparks Amax Coal's rapid growth. (In Coal Age, vol. 79, Oct. 1974, pp. 93-112)
New people, new organization, new policies, new goals have been successfully combined to project the company into a leadership role.

43 New plowing techniques increase Longwall production potential in West German coal mines, by H.W. Wild. (In Min. Cong. J., vol. 55, Dec. 1969, pp. 59-65)
Discusses production, planning and development of the new techniques at Longwall.

44 Pit productivity must beat rising costs, by K. Whitworth. (In World Coal, vol. 1[7], Sept. 1975, p. 27)

Contends that injection of European capital plus National Coal Board drive should keep coal prices competitive with other fuels in United Kingdom.

45 Policies and constraints for major expansion of US coal production and utilization, by W.R. Hibbard. (In Ann. Rev. Energy, vol. 4, 1979, pp. 147-174)
Partial Contents: Capital investment and operating costs in mining; Coal preparation and beneficiation; Coal taxes and royalties.# Partial Tables: 5, Estimated capital costs (1975 dollars/annual ton); 7, US coal tax and royalty.

46 Saving money with coal geophysics, by J.F. Abshier, G.E. McBride and S.F. Beardsmore. (In Coal Age, vol. 84[9], Sept. 1979, pp. 100+)
Discusses new methods that can help to stretch exploration dollars.

Pricing

47 Coal price, by M. Ippolito. (In Rev. de l'Energie, Aug./Sept. 1979, pp. 649+)

48 Coke reclaiming pays off, by J.R. Burger. (In Coal Age, vol. 84(11), Nov. 1979, pp. 144+)
Pittsburgh Coal and Coke recovers a high-priced production from refuse and will use same plant to wash coal.

49 The conversion of coal to gas and oil, by W.L.G. Muir. High Wycombe, Bucks.: Muir Coal Industry Information Service, 1978. 48p. (Progress Report, 4)
Partial Contents: Gas--prices and costs; Oil--prices and costs.

50 We will use coal if the price is right, by A. Hawkins. (In Combustion, vol. 45, Sept. 1973, pp. 13-14)
Contends that coal price increases in UK, together with smaller rises in the price of other power station fuels, have had a crippling impact on electricity costs in recent years, and the electricity industry, with its tariffs artificially restrained by government policy, is now experiencing the hardest financial times in its history.

Statistics

51 Coal utilization, by D.F. Crickmer. (In Min. Eng., vol. 25, Feb. 1973, pp. 89-91)
Tables: Bituminous coal consumption in US, 1967-1975; Analysis of fuel for electric generation ... with composite average costs.

52 Coal utilization, by R.H. Quenon. (In Min. Eng., vol. 23, Feb. 1971, pp. 210-218)
Tables: Fuel required, 1968, 1980, 1990; Generating capacity, 1968, 1980, 1990; Bituminous and coal exports, 1960-1970; Bituminous coal imports, 1960-1970.# Industry activity: Flue gas desulfurization demonstrations--with remarks on status, cost/finance.

53 Japanese steel industry dominates outlook for Western coking coals, by W.B. Beatty. (In Coal Age, vol. 77, Nov. 1972, pp. 88-94)
A review of developing Pacific Coast supplies and a survey of world reserves.# Tables: Western Canada coking coal production estimates, 1972, 1975 and 1980; Western Canada coking coal shipments, 1975, 1980; Principal sources of Japanese coking coal; World reserves of coking coal.

54 Outlook for the HPI: king coal's rebirth, by J.D. Wall. (In Hydrocarbon Proc., vol. 53, May 1974, pp. 89-91)
Tables: 1, Projected world coal production; 2, Principal coal conversion processes; 3, Comparable fuel costs.

55 Top 15 coal producing groups in 1972. (In Coal Age, vol. 78, Apr. 1973, pp. 39)
Production in 1972 totaled 301,208,359 tons against 265,648,585 for the 15 ranking groups in 1971.

56 World metallurgical coal in the seventies, by W. Bellano. (In Min. Cong. J., vol. 57, Feb. 1971, pp. 80-86)
Tables: 1, World steel production with annual rate of growth, 1960-1980; 2, Estimated coking coal reserves in selected countries; 3, World production in oven and beehive coke in selected countries, 1960-1980; 4, World coal and lignite production, 1965-1980; 5, Major international coal movements; 6, USA--origin districts for export coal (thousand metric tons).

57 Watch those costs, by W.F. Saalbach. (In Coal Age, vol. 75, Sept. 1970, pp. 70-74)
Tables include: US population, energy, power and coal data for selected years, 1955, 1965, 1975, 1985; How wholesale price indexes varied from 1957 through 1969; How productivity has increased during the period 1950 through 1967.

II. ELECTRICITY

Costs

58 Are your substation costs too high? (In Elect. World, vol. 173, June 29, 1970, pp. 26-27)
The true installation costs of substations can vary according to the way in which they are purchased. Examination of three purchase plans shows how costs really compare. # Tables: Importance of disconnect costs; Dollar-costs breakdown; Total savings possible, adjusted for engineering costs, are charted for given examples.

59 Automation influences system costs, by R.F. Cook. (In Elect. World, vol. 182, Sept. 15, 1974, pp. 87-89)
Distribution automation holds the potential for reducing operating costs as well as improving system reliability. The industry must accept, however, the philosophy of selective load control.

60 Autotransformers ease cost of raising distribution voltage. (In Elect. World, vol. 173, Feb. 16, 1970, pp. 38-39)
Autotransformers solidly connected to feeders as the interface between distribution circuits of different voltage levels can forestall the large lump sum expenditures usually required to change over a system to a higher voltage.

61 Changing electric utility economics and the role of combustion turbines: abstract, by R.A. Brown. (In Combustion, vol. 47, July 1975, pp. 16-17)
Discusses the impact of changed factors on the electric utility industry and demonstrates that their effects can be lessened through the judicious use of combustion turbines in their open cycle or combined cycle. Also discusses operating expenses. Abstract of paper presented at the American power conference, Chicago, 1975.

62 Combined cycle power plant capital cost estimates. San Francisco: Power Research Institute, 1977. 217p.
Presents capital cost estimates for oil-fueled, high-efficiency combined gas turbine power plants in five regions of the United States. Cost estimates were based on information from construction projects for both conventional and combined cycle plants.

63 Comparative economics of self-generated and purchased power, by P. Caude. (In Chem. Ind., Sept. 6, 1975, pp. 717-722)
Contents: Cost of energy sources; Purchased electricity--primary fuels, generating efficiency, capital charges, tariff and pricing policies.# Partial Charts: Price trends of primary fuels (as 1974 £), 1970-1982; Price trends of primary fuels (as current £), 1970-1982; Generating costs; Generating costs as function of utilization; Price of electricity from public supply 10MW 80 percent.# Table: Price of electricity generated privately, 1977.

64 Computers help power users, by D. Millchamp. (In Engineering vol. 210, July 24, 1970, pp. 88)
By improving the operating efficiency and commissioning time of new power stations, the Central Electricity Generating Board's (CEGB's) hybrid computer could lead to lower power costs.

65 Conceptual design and cost estimate 600 MWe coal-fired fluidized-bed combined cycle power plant, by D.A. Huber and R.M. Costello. (In Combustion, vol. 50, June 1979, pp. 22-28)
Reviews recent studies which indicate that combined cycle power plants, using fluidized-bed combustors, offer the potential for the production of electrical power from coal in an environmentally acceptable manner at a higher efficiency and a lower cost than conventional coal-fired steam power plants incorporating flue gas desulfurization systems.# Partial Tables: Preliminary direct capital cost estimate; Commercial plant capital cost summary, mid-1977 dollars; Commercial plant annual cost, summary cost of electricity; Sensitivity of capital cost estimate of some economic and design criteria; Capital cost comparison.

66 Containing the cost of undergrounding, by D.M. Cherry. (In IEE Proc., vol. 122, Mar. 1975, pp. 293-300)
Analyzes the value of the many significant advances in cable design and application in reducing the cost of cables relative to that of the overhead lines.

67 Control construction with computer, by T.C. Elliott. (In Power, vol. 115, Oct. 1971, pp. 45-49)
New management technique links headquarters computer with field sites to keep loose tabs on construction and cost scheduling.

68 Cost of construction lighting cut. (In Elect. World, vol. 176, Oct. 1, 1971, pp. 67-68)
The Tennessee Valley Authority sidestepped much of the usual cost of temporary construction lighting at its massive Brown Ferry Nuclear Plant, by going directly to the installation of the permanent lighting systems.

69 Cost program recognizes forward outages. (In Elect. World, vol. 175, June 1, 1971, pp. 68-69)

System production costs are determined by a new computer program, which includes the cost penalty effects of forced outages in studies of systems with thermal and hydro units.# Chart: Total expected system costs.

70 Cost separation of steam and electricity for a dual-purpose power station, by P. Leang. (In Combustion, vol. 44, Feb. 1973, pp. 6-14)
Discusses several methods of separating the costs of steam and electricity. These methods are developed in conformance with thermodynamic and economic principles.# Partial Tables: Economic criteria; Energy equivalence method of fuel cost allocation; Fuel cost allocation to steam based on an established electricity cost; Capital cost separation for steam and electricity; Capital cost allocation by cost separation of major functions; Cost separation of joint components; Capital cost allocation based on the single-purpose electric generating plant capital cost; Allocation of operation and maintenance costs; Unit cost based on fixed annual capacity factor; Unit cost based on fixed peak demand; Cost allocation for partial load operation.# Charts: Energy equivalence method of fuel cost allocation; Established electricity cost method of fuel cost allocation; Alternative cost justifiable investment method.

71 Cost study of super heating geothermal steam, by J.M. Hensler and R.C. Axtmana. (In Energy, vol. 4[3], June 1979, pp. 365-371)
Compares the electrical generating costs for hybrid plants in which fossil fuel super-heats geothermal steam with those for conventional dual-flash plants. Gives design and cost analysis.

72 Costs look promising for north-central power grid. (In Elect. World, vol. 177, Jan. 1, 1972, pp. 42-43)
Initial economics look promising for development of vast coal resources in the north-central United States through a multibillion-dollar system of large mine-mouth plants and EHV lines to supply up to 53,000 Mvv to load centers across that area.

73 Costs of electricity supply interruptions to industrial consumers, by A.F. Jackson and B. Salvage. (In IEEE Proc., vol. 121, Dec. 1974, pp. 1575-1576)
Presents the results of a survey into the effects of costs of electricity supply interruptions to 23 large industrial consumers. Detailed costs have been obtained for 12 companies and it is shown that they can be expressed in terms of a fixed cost, varying with the frequency of interruption, and a variable cost, varying with the duration of interruption.

74 Cut cost with integrated coal handling, by R.L. Longfellow. (In Power Eng., vol. 76, Aug. 1972, pp. 30-33)

Total systems approach to unloading, stockpiling and reclaiming has led to cost reduction of up to 50 percent compared with former methods of handling water-transported coal.

75 Dimmers: cost savings and environmental control flexibility, by L. Kellerman. (In Lighting Design Appl., vol. 2, Oct. 1972, pp. 53-55)
Partial Contents: Cost savings estimate; Prominent buildings; Large meeting halls.# Tables: Comparison between dimming system and switch control; Features and comparative characteristics of various types of light dimmers.

76 Distribution spending reflects growth. (In Elect. World, vol. 175, Mar. 15, 1971, pp. 49-53)
Upturn in underground boosts distribution spending while transmission slowdown cuts the per-kva cost of substations.# Tables: Expenditure for lines and substations, 1970, 1971; New lines and substations energized.

77 Economic allocation of regulating margin, by W.O. Stadlin. (In IEEE Trans. Power Appar. Syst., vol. 90, July 1971, pp. 1776-1781)
A coordinating technique is determined which determines the least expensive generation schedule required to maintain a total specified power system regulating margin, where a distinction is made between spinning reserve and regulating margin. The effect on operating cost of applying the regulating margin restraint is also investigated.

78-79 Economic analysis of a generalized design for a forced cooled cable, by H.M. Long, and others. (In IEEE Trans. Power Appar. Syst., vol. 90, May 1971, pp. 1232-1243; Discussion, pp. 1244-1245)
Paper presents part of the results of a study conducted under the sponsorship of the Edison Electric Institute with three broad objectives: 1, To determine which forced cooling techniques are readily applicable to state-of-the-art high pressure, oil filled, pipe-type cable lines; 2, To prepare a generic design of a typical cooling system for specimen pipe-type line; 3, To establish the economic advantage of the application of forced cooling to this cable system. This generic design is used as the basis for an economic analysis.

80 Economic awareness in design planning, by C.L. Amick. (In Lighting Design Appl., vol. 3, Jan. 1973, pp. 16-21)
Reasserts the importance of useful cost analysis, and suggests practical guidelines in system and production comparisons, and introduces readers to new and accepted cost evaluation methods.# Tables: 1 & 2, Lighting cost analysis--basic data, initial costs, operating and total annual costs; 3, Effect of energy rate.

81 Economic comparisons in planning for electricity supply, by

T.E. Norris. (In IEEE Proc., vol. 117, Mar. 1970, pp. 593-605; Correction, vol. 118, Mar./Apr. 1971, pp. 570; Discussion, vol. 117/118, Sept. 1970, July 1971, pp. 855-856, 905-906)
Outlines general principles for setting up a comparison of alternatives, and describes the single basic concept for economic comparisons, namely, present valuing. Present valuing (or discounting) is essential to take account of the timing of costs and revenues (or cash flows). Three equally acceptable methods of comparison, commonly known as discounted cash flow, present value and annual cost methods, are described and compared.

82 The economic cost of power system reliability, by N.D. Uri. (In Energy Comm., vol. 4[3], 1978, pp. 257-269)
Estimates the cost to consumers of providing an additional unit of generating capacity built on a loss of load probability criterion of 0.13 days per year.

83 Economic study favours 34.5kv., by E.D. McCollon. (In Elect. World, vol. 183, May 15, 1975, pp. 54-57)
Distribution engineers must reevaluate higher voltage systems to meet future loads. Load growth and changing patterns of energy use may call for conversion or two-voltage overlay.# Tables: 1, Investment requirement (millions of dollars); 2, Annual costs (thousands of dollars); 3, Load density.# Distribution investment has decreased 2.3 percent annually; Investment requirements are affected by firm capacity; Costs drop sharply as load density reaches 12Mva/sq. mi.; Revenue needs in the expansion area vary linearly; Revenue requirements in new areas--peak is about 10 years.

84 Economics of unit substations, by T. Kociuba. (In Elect. Const. Maint., vol. 71, June 1972, pp. 76-77)
Transformers with higher-than-standard impedance can make significant economies possible in the cost of secondary sub. Gives sample calculations and a table showing cost comparison.

85 Electric motors versus gas engines as prime movers in producing operations, by J.K. Ballou. (In IEEE Trans. Ind. Gen. Appl., vol. 6, Nov. 1970, pp. 540-544)
Of the major problems in the design of oil production facilities is the selection of prime movers for power-driven equipment required in the production and transportation of crude oil. The economic reasons behind the trend toward more widespread use of electric motors and away from the historical use of gas engines for the large percentage of installations are summarized.# Charts: 1, Fuel or power input versus power output; 2, Gas engine operating cost; 3, Electric-power operating cost; 4, Comparison: motor and engine-operating cost versus horsepower.# Table shows water flood pump cost analysis.

86 Electricity supply distribution: trimming for efficiency, by K.G.W. Bolton. (In Electron. Power, vol. 16, Apr. 1970, pp. 124-127)

Contends that area electricity boards can only influence a small proportion of their total operating costs by improving operating efficiency. Ways of improving efficiency include better use of manpower, use of computers for dealing with economic and technical matters, lowering of engineering costs by increased standardization and exploitation of new materials and techniques....# Charts: 1, YEB proportionate operating costs for financial year 1968-69; 2, YEB operating costs relative to 1959; 3, YEB productivity and capital expenditure.

87 Estimating electric power consumption in oil-producing operations, by R.N. Rhodes. (In IEEE Trans. Ind. Gen. Appl., vol. 6, July 1970, pp. 364-374)

Electric power consumption for economic studies of oil-producing operations is generally overestimated by 30 to 70 percent. Methods for making more accurate estimates are given, and test data that will be helpful when detailed information is not available are provided.

88 Fluorescents on/off. (In Lighting Design Appl., vol. 3, Jan. 1973, pp. 38-39)

Presents cost formula for calculating.

89 Forecasting minimum production costs with linear programming, by J.T. Day. (In IEEE Trans. Power Appar. Syst., vol. 90, Mar. 1971, pp. 814-821; Discussion, pp. 822-823)

Presents an approach to and the results of applying linear programming to the forecasting of minimum production costs of electric utility systems. Describes the linear problem format used in the optimization of the start-up and load allocation of generating units. Included are results from investigations of the influence of linear versus nonlinear unit cost/output curves and the number of time segments used to approximate a load duration curve. A comparison is made of the production costs obtained from an established program. The speed, accuracy and flexibility achieved by using linear programming techniques make it a desirable method for production costing for many study purposes.

90 Generation in focus: trends and issues in electric power production--Benefit cost ratio, by J. Papamarcos. (In Power Eng., vol. 76, June 1972, p. 20)

Discusses the preparation of benefit cost analysis.

91 How much does a lighting system cost? by W.B. DeLaney. (In Lighting Design Appl., vol. 3, Jan. 1973, pp. 22-28)

Partial Contents: Discounted cash flow explained; Estimating corporate rate or return; Income tax effects; Depreciation of equipment; Operating expenses.# Partial Charts: Cash flow for a $10,000 investment ...; Long-term capitalization; Tax effect on income; Cost analysis--Basic data, Initial costs, Operating costs.

92 How to fight rising construction cost: panel discussion. (In Elect. World, vol. 174, Dec. 1, 1970, pp. 36-38)
Discusses, among other topics, building costs, and the problem of overtime.

93 Incremental voltage uprating of transmission lines, by D.F. Shankle. (In IEEE Trans. Power Appar. Syst., vol. 90, July 1971, pp. 1791-1795)
Discusses the technical and economic considerations of incrementally uprating the voltage of existing lines. Cost comparisons are made with building new lines on new right-of-way (ROW).# Tables: 1, Typical transmission line characteristics; 2, Cost alternatives.

94 Industrial combined heat and power: a case history, by J.D. Garney. Ove Arup & Partners, 1978. 27p. illus.
Focuses attention on the anticipated large-scale savings which appear to be achievable from major schemes involving on-site generation of electricity.# Partial Contents: Principles of feasibility--open options and an enquiring attitude, capital costs, operating costs; Procedures and methods--costs, gross and net annual savings, economic viability.# Tables: Typical annual operating schedules; Net annual savings as a percentage of estimated current fuel costs at 1976 prices; Light engineering complex capital costs and annual savings (Dec. 1976 prices); Heavy engineering factory capital costs, fuel costs, and savings (1978).

95 Influence of conductor designs and operating temperatures on economics of overhead lines, by P. Dey, and other. (In IEEE Proc., vol. 118, Mar./Apr. 1971, pp. 573-590; Discussion, vol. 188, Sept. 1971, pp. 1285-1286)
Evaluation of capital costs of lines and specific transmission costs have been carried out for the above designs. The interrelationships among the conductor size, the load transmitted, the conductor temperature and the specific transmission costs is shown for selected loads.

96 Infrared promotion seeds heating market. (In Elect. World, vol. 176, Nov. 1, 1971, pp. 88-89)
Utilities find packaged infrared heating promotion serves to stimulate sales reps and customers' interests into action.

97 Is power to the people going underground? by G.D. Friedlander. (In IEEE Spect., vol. 9, Feb. 1972, pp. 62-71)
A report on an important conference regarding the economic feasibility of and technological progress toward the construction of underground high voltage transmission lines.

Includes table on capital costs of 3000-MVA, 132-kv cryogenic cables, percent.

98 Light system cost comparison, by J.F. Finn. (In Lighting Design Appl., vol. 5, Jan. 1975, pp. 26-27)
Contends that there are a number of options that require professional analysis to insure the efficient use of energy. Presents calculations and a table showing lighting system cost comparison.

99 Liquid heat from thermal fluid lowers long-term expenses, by C. Beatson. (In Engineer, vol. 236, Mar. 1, 1973, pp. 34-35)
Contends that British industry has been slow to adopt a method of heating which offers greater efficiency and cheaper running costs than other systems.

100 Low cost power myth, by W.J. McCarthy and R.L. DeMumbrum. (In Chem. Eng. Prog., vol. 68, Mar. 1972, pp. 27-30)
Looking ahead three decades the cost of electricity to large industrial customers will almost double. The cost of fossil fuels will increase $2\frac{1}{2}$ to 3 times.# Contents: Charting future costs; What's happening to utility costs; Nuclear capacity reduces fuel costs; The rate to industry.# Charts: 1, System load forecast for Southeastern Michigan; 2, Turnover ratio is an indicator of plant investment; 3, Average fuel costs will drop under the influence of nuclear fuel; 4, Breeder reactors will result in lower system heat rate; 5, Cost components of the industrial rate structure will change significantly; 6, Forecast of the average industrial rate in Detroit Edison system.

101 The marginal cost of electricity used as backup for solar hot water system: a case study, by R. Bright and H. Davitian. (In Energy, vol. 4[4], Aug. 1979, pp. 645-661)
Presents a method developed for estimating the long-run marginal cost to electric utilities of providing backup service for solar residential heating and hot water (HHW) system. Accounts for all investment, fuel, and operating costs required to provide the added electric service for HHW.

102 Meeting environmental needs: investment costs. (In Combustion, vol. 43, Oct. 1971, pp. 8-9)
An estimate of investment and energy costs, for thermal electric plants scheduled for operation in 1976, clearly demonstrates the incremental costs associated with power plant environmental requirements. The projected costs include escalation, and apply to four types of representative 800 and 1100 Mw thermal electric plants.

103 Meeting environmental needs: water resources, by W.D. Patterson. (In Combustion, vol. 43, Oct. 1971, pp. 9-12)

Contends that the growing demand for electric power and the increased emphasis on reliability require continued planning, design and construction of large new generating facilities. Delays coupled with the design and operating changes often required, are rapidly increasing plant investment and operating costs.

104 Must HVDC go to UHV levels? by P. Danfors and B. Hammarkund. (In Elect. World, vol. 179, Apr. 1, 1973, pp. 60-61)

Selecting a proper voltage level for high-capacity transmission lines is complex. System planners must access use of UHV transmission for their future systems--or as an alternative, HVDC. The economic design of both ac and dc transmission lines and their power carrying capacity is discussed.# Tables: Bipolar HVDC costs--do not include right-of-way costs, but do include capitalized losses. Costs are in Swedish kroner per km per kw; Power capability based solely on economic design.

105 Net energy analysis of electrical transmission lines, by D. Mau and others. (In Can. Elect. Eng. J., vol. 3[4], Oct. 1978, pp. 9-13)

Presents an analysis of the energy costs inherent in construction and operation of a 1000-mile length of 765KV AC, ±400KV DC and ±600KV DC transmission lines.

106 New era of power supply economics, by F.M. Swengel. (In Power Eng., vol. 74, Mar. 1970, pp. 30-38)

Upward trend of equipment prices and construction costs is so widely known and recognized that it is automatically taken into account in power supply planning.# Partial Tables: Changing patterns of generation production costs; Fossil fuel costs required for break-even with nuclear power for a range of fixed charges; Break-even fossil fuel costs; Comparative cost table; Total cost to own and operate; Total annual dollar per kilowatt of capacity with a fuel economy based on oil; Total cost of energy; Excess cost allowable over bus bar cost at central station for elimination of transmission.

107 New generation production cost program to recognize forced outages, by M.A. Sager and others. (In IEEE Trans. Power Appar. Syst., vol. 91, Sept. 1972, pp. 2114-2120; Discussion, pp. 2121-2123; Reply, pp. 2123-2124)

Describes a newly developed computer program designed to analyze the generation production costs to be expected when systems contain hydro-electric plants where energy availability can only be described in probabilistic terms and hydrothermal (or all thermal) generating systems where the forced outages of large units can cause considerable added production expense and high levels of operation of peaking unit.

108 On-line costing system aids construction planning: cost/general, by R. Parsley and G. D. Rinehart. (In Elect. World, vol. 183, Mar. 15, 1975, pp. 106)
The installation of cost/general, an on-line computer-based management control system, at San Diego Gas & Electric Co. has provided some important advantages--particularly for the utility's large construction commitments.

109 On the spot power generation evens out peaks and troughs, by C. Beatson. (In Engineer, vol. 235, July 6, 1972, pp. 46-47, 49, 51)
The effect of peak lopping on the chargeable minimum demand for a typical year's operation is shown in a graph. Investment cost including spares is also shown.

110 Operating costs jump 20% in two years. (In Elect. World, vol. 180, Nov. 1, 1973, pp. 43-46)
Provides operating data for 43 modern stations containing 34,888Mw of generating capability.# Tables: Fuel exceeds 79 percent of operating cost in all but smallest group; Manpower requirements reduced for large stations; Outdoor construction shaves investment; Design for burning gas slashes station investment.# Included is the 18th steam station cost survey.

111 Operating lamps for maximum economy: question and answers. (In Elect. Const. Maint., vol. 71, Oct. 1972, p. 112)
Answer to question about increased cost of electricity resulting from leaving lamps on 24 hours a day; involves calculations which show energy cost.

112 Optimal generation planning considering uncertainty, by R. R. Booth. (In IEEE Trans. Power Appar. Syst., vol. 91, Jan. 1972, pp. 70-77)
Describes a method of production costing based on probabilistic simulation methods, combined with an advanced dynamic programming formulation of the problem in order to treat uncertainty in a systematic manner. Some computational features are described and examples given of the use of the procedure on a practical system.

113 Optimal real and reactive power operation in a hydrothermal system, by R. Billinton and S. S. Sachdeva. (In IEEE Trans. Power Appar. Syst., vol. 91, July 1972, pp. 1405-1411)
Discusses the system state as defined by the system variables. The input-output function of a mixed hydrothermal system consists of linear (hydro input costs) and non-linear (thermal input costs) functions. Table five is a comparative study of real power cost for the three optimation considerations.

114 Optimum power flow for systems with area interchange controls,

by J. Peschon. (In IEEE Trans. Power Appar. Syst., vol. 91, May 1972, pp. 898-903; Discussion, pp. 904-905; Reply, pp. 904)

Minimizes the hourly operating cost of an interconnected power system constrained by prescribed area interchanges. The scheduled area interchanges act as constraining equations in the economic hourly optimization procedure. The cost minimization is performed with respect to active and reactive generation and is subjected to constraints on voltage and power flows.

115 Planning future electrical generation capacity: a decision analysis of the costs of over- and under-building in the US Pacific Northwest, by A.P. Sanghvi and D.R. Limage. (In Energy Policy, vol. 7[2], June 1979, pp. 102-116)

Develops a decision analysis framework to study the needs for additional electrical generating capacity in the presence of divergent load growth forecasts. A method for determining the costs of over- and under-building capacity is developed. The results support the conclusion that in the Pacific Northwest the social costs of over-building are lower than the costs of under-building.# Tables: 1, Payoff matrix showing electricity costs in 1988, assuming a growth scenario in which load growth outcome is discovered in 1985 (1978 mils/kWh); 2, Payoff matrix showing electricity costs in 1988, assuming a growth scenario in which load growth is discovered in 1988.

116 Plant capital costs spiralling upward. (In Elect. World, vol. 176, July 1, 1971, pp. 36-38)

Escalation trends, contingency requirements of a given site or project, and the duration of overall construction schedules, are said to be the three main factors governing future costs of new generating plants.# Tables: 1, Capital cost of generating plants, 1970, 1980; 2, Annual escalation rates; 3, Relative cost of field labor; 4, Plant cost contingencies; 5, Typical construction schedules; 6, Trend of plant costs, 1970-1980.

117 Pole treatment can pay dividends. (In Elect. World, vol. 174, Oct. 1, 1970, pp. 47-48)

Tables: 1, Comparative costs of treatment vs. replacement...; 2, Comparative costs of treatment vs. replacement, 9 percent interest and straight-line depreciation.

118 Power distribution costs cut, by E. Jeffs. (In Engineering, vol. 208, Sept. 19, 1969, p. 310)

The North West Electricity Board has saved £7M in the last three years through the use of new engineering techniques and revision of plant specifications. Big savings have been made by switching to aluminum cable and standardizing on transformers.

119 Power plant capital costs going out of sight, by F.C. Olds. (In Power Eng., vol. 78, Aug. 1974, pp. 36-43)
Projections of power plant costs have been escalating at the rate of about 26 percent annually since 1970 and they still appear to be underestimating the total cost up to the time of commissioning; while the difficulty of estimating continues to increase, the need for more accurate estimates becomes ever more urgent.# Partial Contents: Cost picture in 1965-66; Early cost analysis; Coming to grips with higher costs; Establishing a new base cost; Determining a new base cost; Fossil plant cost projections.# Partial Charts: 1, Cost per KW for new fossil plants; 2, Range of nuclear plant costs in dollars/KW for all plants contracted for each year, 1965-74--costs shown are those estimated by owner in the year the contract was let; 3, Analysis of cost trend for AEC's reference 1000-MWE nuclear plant; 4, Range of nuclear plant costs in dollars ... year of plant announcement/generally year end; 5, Average cost ... legend ...; 6, Millions of dollars ... legend ...; 7, Fossil-nuclear estimated cost comparison--direct cost only.# Partial Tables: Nuclear plant costs and schedules; Nuclear plant costs schedules, 1974; Loading on direct cost estimates for coal and nuclear plant.

120 Power reliability cost vs. worth, by R.B. Shipley and others. (In IEEE Trans. Power Appar. Syst., vol. 91, Sept. 1972, pp. 2204-2207; Discussion, pp. 2207-2212)
Presents a method for determining optimum service reliability. A method of estimating the cost of service interruptions is proposed and the cost obtained is compared with values determined by others. The intention of this paper is to provoke discussion and further research into the cost of power interruptions and the optimizing of system design considering the cost.

121 Power system production cost calculations: sample studies recognizing forced outages, by M.A. Sayer and A.J. Wood. (In IEEE Trans. Power Appar. Syst., vol. 92, Jan. 1973, pp. 154-158)
Presents results of energy production cost calculations made for the three power systems using a new probabilistic technique which computes the statistical, expected value of the energy cost. Paper shows quantitatively the major importance of unit forced outages on calculated expected power system energy costs. The energy cost penalty effect ranged from 2 to 5 percent of the total system energy cost for the sample systems studied.

122 Present practices in the economic operation of power systems (In IEEE Trans. Power Appar. Syst., vol. 90, July 1971, pp. 1768-1774; Discussions, pp. 1774-1775)
Presents practices within the utility industry in the economic operation of electric power systems.# Contents:

Determination of basic cost data--incremental costs, unit input-output curves; other plant data, dispatch fuel price/ water value or worth, incremental maintenance costs, transmission losses; Economic dispatch--on-line economic dispatch; Interconnected operations.

123 Projected cost of energy, by J.A. Lane. (In Combustion, vol. 42, Dec. 1970, pp. 34-39)
An appraisal of the future outlook for low-cost nuclear power in the light of economic changes occurring since 1967.# Contents: Nuclear plant costs; Definition of low-cost power; Transmission and distribution costs; Reducing nuclear power costs.# Charts: 1, Unit power costs vs. initial capacity; 2, Unit capital cost vs. date of operation; 3, Per capita consumption of electricity in the US; 4, Trends in unit size of turbine generator; 5, Past and projected costs of nuclear plants.# Tables: 1, Cost of 500 Mw(e) Niagara Mohawk nine mile point; 2, Estimated nuclear plant costs; 3, Weighted average costs; 4, Present and potential electricity use in the United States; 5, Break-even cost of electricity required to replace gas heat loads; 6, Characteristics of mixed fuel cycle unclad metal breeder reactor.

124 Reduce operating costs through maintenance, by T.W. Abolin and H. Ferguson. (In Power, vol. 116, May 1972, pp. 93-95)
Program calls for checking heat rate of plant equipment periodically and correcting deficiencies in the most inefficient components first.

125 Reducing the cost of electricity supply, by A.S. Windett. Gower, 1980. 144p.
A working tool for the industrial and commercial user of electricity interested in paying the minimum for this power supply.

126 Research helps cut Britain's power costs. (In Power, vol. 113, Dec. 1979, pp. 60-61)
Central Electricity Generating Board (CEGB), prime customer for British power-generation equipment, carries out major research programs. This is a glimpse at its non-nuclear R&D program.

127 Residential electric heating and cooling: total cost service, by J.G. Asbury, R.F. Giese and R.O. Mueller. Illinois: Argonne National Lab., 1979. 19p.

128 Resource optimization and economic planning, by J.W. Griffith. (In Lighting Des. Appl., vol. 3, Sept. 1973, pp. 23-27)
Economic models and life-cycle costing are more desirable than rules of thumb in establishing design tradeoffs for energy utilization.# Table: Two economic models compared.

129 Saving justifies capacitors on underground feeders. (In Elect. World, vol. 173, June 22, 1970, pp. 36-37)
Studies show that capacitors are economically attractive even on all underground feeders; combination of one fixed and one switched bank proves to be most economic choice.# Table: Incremental installation cost.

130 Simplify power plant cost calculations, by R.T. Anderson. (In Power, vol. 114, July 1970, pp. 39-41)
Monographs quickly pinpoint fixed, fuel, operating and overall electric generation costs for power plant of 100-2000MW capacity.# Tables: 1, Select the desired electric generating capacity ...; 2, Converting the heating values of coal, oil and gas to energy cost; 3, Find operation and maintenance costs; 4, Surveying fuel costs and heating values nationwide.

131 Small savings count in electricity, by E. Jeffs. (In Engineering, vol. 208, Aug. 1, 1969, p. 107)
Working towards improved operating economy of present and future nuclear power stations, CEGB's Berkeley laboratory has already found ways to save money in both nuclear and conventional stations.

132 Social costs of producing electric power from coal: a first order calculation, by M.G. Morgan and others. (In IEEE Proc., vol. 61, July 1973, pp. 1431-1442; Discussion, vol. 62, July 1974, pp. 1026-1028)

133 Superconductive DC transmission lines: design study and cost estimates, by C.N. Carter. (In Cryogenics, vol. 13, Apr. 1973, pp. 207-215)
Considers capital cost and presents a table showing capital costs in £/MVA km and cost breakdown as percentage of total.

134 System maintenance costs soar. (In Elect. World, vol. 175, Mar. 15, Mar. 1971, p. 60)
US utility appropriations average 25.2 percent over 1969 charges as thin reserve in some regions spurs premium time repairs.

135 Tension-point supports cut overhead line cost. (In Elect. World, vol. 173, Jan. 12, 1970, pp. 30-31)
Designed for Indian 66-kv line, structures with clearance for longitudinal swings prove cheaper than deadends in hilly areas.

136 Transmission line losses based on conductor characteristics, by E. Hazan. (In Power Eng., vol. 73, Sept. 1969, pp. 42-43)
Presents a method for determining operating costs based on conductor characteristics.

137 Trends in capital costs of generating plants, by K.A. Roe and W.H. Young. (In Power Eng., vol. 76, June 1972, pp. 40-43)
Continuation of rapid growth in capital costs will make financing and provision of increases in generating capacity even more difficult to achieve. Effective counter-measures are needed.# Contents: Construction labor costs; Stabilizing construction wages; pushing costs up.# Tables: Handy-Whitham cost index, 1960-1971; Dollar and cents per hour, 1962-1970; Percent increase plant total cost; Capital cost ranges for 800 MWE nuclear generating plants; Capital cost ranges for 800 MWE fossil generating plants, 1965-1980.

138 Underground to overhead cost ratio shows increase, by W.C. Hayes. (In Elect. World, vol. 176, Oct. 1, 1971, pp. 48-50)
Data on cost of URD per front foot that obtains in the companies surveyed.# Tables: Average cost--Far West, Northeast, Midwest, Southeast, for pre 1960, end of 1963, end of 1967, end of 1970, est. 1972, est. 1975; Average cost ratio, underground to overhead--Far West, Northeast, Mid-West Southeast....

139 Unit residential transformer gains as the cost of money rises. (In Elect. World, vol. 173, Mar. 23, 1970, pp. 26-27)
Study shows that higher money cost and type of growth pattern experienced within a service area both strongly influence the comparative economics of URT systems.# Tables: Economic break-even points for URT vs. primary secondary system for various premiums.# Comparative costs are plotted for three different growth patterns and three different levels of carrying charges within each of the growth patterns.

140 Unpaid costs of electrical energy: health and environmental impacts from coal and nuclear power, by W. Ramsey. Baltimore: Johns Hopkins, 1979. 180p.

141 Utility cost study: Pt. 1, Gas turbine utility plant. (In Turbomach. Int., vol. 19[5], July/Aug. 1978, pp. 33-45)
A report of construction cost and production expenses for 1977 of a number of electric utility companies within the United States.

142 Valence of electric energy: a fundamental approach to economic generation planning, by H. Stephenson. (In IEEE Trans. Power Appar. Syst., vol. 92, Jan. 1973, pp. 248-253)
Presents an easy-to-use method for evaluation of energy from power plants with different output and cost characteristics. It is based on the comparison of generation cost of power station to the cost of generating energy with constant power, expressed by a value factor and an analysis of the demand arranged to certain groups of daily utilization hours.

143 What is the true cost of electric power from a cogeneration plant? by W.H. Costois. (In Combustion, vol. 50, Sept. 1978, pp. 8-14)
Contents: Cogeneration economics; Economic characteristics.# Charts: Cost allocation for cogeneration plant products; Cogeneration plant product costs as determined by OSW: The effect of financing method on the economic characteristics of cogeneration plants.

Finance

144 Application of the minimum revenue requirement investment opportunities theory to economic evaluations with complex cash flows, by J.H. Sosinski and others. New York: IEEE, 1977. 8p. (Pap. F78 094-5)
Presents a method for performing economic comparisons of alternatives in the regulatory utility environment.

145 Case for leasing, by P.J. McTague. (In Elect. World, vol. 174, Oct. 1, 1970, pp. 89-90; Oct. 15, 1970, pp. 135-136; Nov. 1, 1970, pp. 105-106; Nov. 15, 1970, pp. 147-148)
This is a four-part article. In part one, the author discusses the general advantages of large-scale leasing; in part two, he goes into the specific economic advantages; in part three, he surveys, in depth, the contractual factors involved in leasing; and in part four, he continues his in-depth investigation.

146 Cash crunch dims power prospects, by E. Rubinstein and G.D. Friedlander. (In IEEE Spectrum, vol. 12, Mar. 1975, pp. 40-44)
Money doesn't buy what it used to, an inflationary fact that's set back the utilities' expansion programs by years.# Presents the financial woes of the US power industry based on the price book ratios of 90 electric utilities listed on the New York exchange in June 1974.

147 Cash crunch: its ripple effect; how the financial difficulties of US electric utilities threaten to erode diverse segments of the economy, by E. Rubinstein. (In IEEE Spectrum, vol. 12, May 1975, pp. 41-43)
Contents: Facing facts in Michigan; Reliability over power; Energy intensive industries; The utility suppliers.

148 Corporate model programs for system planning evaluations, by M.A. Sager and A.J. Wood. (In IEEE Trans. Power Appar. Syst., vol. 91, May 1972, pp. 1079-1084)
Utility corporate models have been developed to provide financial planning assistance in many utilities. Paper describes new corporate models designed specifically for assisting system planners in performing economic and financial evaluations of alternate plans. Programs described from an

integrated data processing and evaluation system to produce utility corporate models for both investor-owned and government-sponsored electric utilities. Paper concludes that models for planning evaluations would not seem to require monthly financial models; but there is no reason why models designed to study short-term cash flows cannot be used for planning evaluations.

149 Corporate models: a new tool for planning, by A. J. Wood. (In IEEE Trans. Power Appar. Syst., vol. 90, Mar. 1971, pp. 401-408)
Discusses the role of corporate models in the planning of the growth and operation of electric utilities. Also, reviews briefly the contents of these models for electric utilities and discusses their relationship to the more conventional approaches to engineering economy and financial planning studies.# Partial Figures: Cash flow pattern for a particular utility; Logical outline of an interactive, financial planning program; Example of an income statement from an interactive corporate model--data pertain to the retirement study case.

150 Electrical World annual statistical report, 1971: capital expenditures. (In Elect. World, vol. 175, Mar. 15, 1971, pp. 42-48)
Tables: Total capital expenditures; Generation capital spending; Transmission capital spending; Total electric power system capital expenditure (millions of dollars); Electric utility capital spending, 1970-1971; Electric power system spending by ownership (thousands of dollars); Total power system budgets, 1968-72 (millions of dollars); Electric utility capital spending by regions (thousands of dollars).

151 Electrical World annual statistical report, 1971: financial. (In Elect. World, vol. 175, Mar. 15, 1971, pp. 63-66)
Revenues, net income hit record in 1970.# Tables: Security sales by investor-owned electric utilities (thousands of dollars; Investor-owned electric utilities combined balance sheets (millions of dollars); Distribution of the electric revenue dollars (percent); Investor-owned electric utilities income statement (millions of dollars).

152 Electrical World annual statistical report, 1972: capital expenditures. (In Elect. World, vol. 177, Mar. 15, 1972, pp. 44-50)
Record spending in 1971 to be eclipsed by cost of massive expansion program.# Tables: Total electric power system capital expenditures (millions of dollars), 1961-1972; Total capital expenditure, 1971-1972; Electric utility capital spending, 1971-1972; Electric power system spending by ownership (thousands of dollars), 1971-1972; Total power system budgets, 1969-73 (millions of dollars); Bulk capital spending falls in three fast-growing regions, 1971-73; Electric utility capital by regions (thousands of dollars), 1971-1972.

153 Electrical World annual statistical report, 1973: capital expenditures. (In Elect. World, vol. 179, Mar. 15, 1973, pp. 38-44)
Contents: US utilities boost budgets by $2.7 billion for 1973; East North Central region regains lead in 1973 budgets. # Partial Tables: Total electric power system capital expenditure (millions of dollars), 1962-1972; Total capital expenditure; Electric utility capital spending (thousands of dollars), 1972-73; Total budgets, 1970-74 (million dollars); Bulk capital spending falls in three fast-growing regions.

154 Electrical World annual statistical report, 1973: financial. (In Elect. World, vol. 179, Mar. 1973, pp. 60-62)
Total electric revenues of the entire utility industry climbed to a record $27.9 billion for 1972. Investor-owned utilities collected 86.2 percent of revenue. # Tables: Investor-owned electric utility income; Electric rate incomes, 1962-72; Distribution of electric revenue, 1968-72; Security sales by investor-owned electric utilities (thousands of dollars), 1962-1972; Investor-owned electric utilities combined balance sheets (millions of dollars), 1968-72.

155 Electrical World annual statistical report, 1975: capital expenditures. (In Elect. World, vol. 183, Mar. 15, 1975, pp. 47-52)
Spending sets new record budgets, foretell 4.6 percent boost in 1975. # Tables: Total electric power system capital expenditures (millions of dollars), 1964-1975; Electric utility capital spending (thousands of dollars), 1974-1975; Heaviest capital spending falls in fast-growing regions.

156 Electrical World annual statistical report, 1975: financial. (In Elect. World, vol. 183, Mar. 15, 1975, pp. 68-70)
Rate boosts and fuel adjustments ease financial problems slightly. Electric revenues for the entire industry set a new record of slightly over $39 billion in 1974, 23.2 percent above the previous record set in 1973. # Partial Tables: Investor-owned electric utilities combined balance sheet (millions of dollars)

157 Electrical World executive conference, 4th. (In Elect. World, vol. 176, Dec. 1, 1971, pp. 25-26)
Discussed financial problems and came up with many solutions. Over 140 financial experts from electric utilities, banks, investment firms and allied institutions took part.

158 Electricity and natural gas rate issues, by C.J. Cicchetti and M. Reinbergs. (In Ann. Rev. Energy, vol. 4, 1979, pp. 231-258)
Contents: Is marginal cost pricing still an issue?; Time-of-use pricing versus load management; Determination of marginal capacity costs; The determination of marginal costs of transmission and distribution facilities.

159 Financial statistics of electric utilities and interstate natural gas pipeline companies. Washington: Department of Energy, Energy Information Administration, 1978. 19p.
Presents financial statistics for privately owned electric utilities and interstate natural gas pipeline companies.

160 Incorporating investment opportunities into economic evaluations to minimize corporate revenue requirements, by James H. Sosinski and others. (In IEEE Trans. Power Appar. Syst., vol. 5, PAS-97(1), Jan./Feb. 1978, pp. 251-260)

161 Power system project priority index formulations, by S.B. Dhar. (In IEEE Power Eng. Sec., Winter meet., 1978. 9p.)
Presents a systematic procedure developed for calculating bulk power transmission project Priority Index Number (PIN), Reliability Improvement Factor (RIF) and Capital Investment Ratio (CIR).

162 Rate increases growth raise profits, by P.R. Jewell and R.M. Sigley. (In Elect. World, vol. 182, Sept. 15, 1974, pp. 126-128)
Computer program shows that rate increases and growth in kw/hrs are most significant determinants of profitability. Increasing costs and interest rates reduce profitability but effect is less than expected. # Tables: Effect of load growth on earnings per share shows sensitivity of profitability to load factor with different growth assumptions; Effect of rate increase on earnings; Composite projected to 1979, 5 percent rate increase/year and 8 percent inflation in labor and materials; Effect of rate increases showing increased costs.

163 [Transmission and Distribution] budgets jump to $6.15 billion. (In Elect. World, vol. 177, Mar. 1972, pp. 51-55)
Contents: Transmission expenditures; Distribution budgets; Substation budgets. # Tables: Expenditures for lines and substations, 1971-1972; New lines and substations energized, 1971-1972.

164 [Transmission and Distribution] budgets up 20 percent over 1972 spending. (In Elect. World, vol. 179, Mar. 15, 1973, pp. 45-49)
Contents: Distribution budgets; Substation expenditures. # Tables: Expenditures for lines and substation, 1972-1973; New lines and substations energized, 1972-73.

165 Utilities eye insurance field for a new source of funds, by R.A. Stout. (In Elect. World, vol. 179, Feb. 15, 1973, pp. 28-29)
Discusses the intention of US electric and combination utilities to endorse, for profit, a low-cost supplemental

health insurance program, to be offered to their customers. # Table: Estimated profit for utility-sponsored insurance (per 500,000 customers).

166 Ways around the cash crunch, by G. D. Friedlander and E. Rubinstein. (In IEEE Spectrum, vol. 12, Apr. 1975, pp. 62-65)

Presents options for retracking the US utilities onto their traditional path of adequate and reliable supply. Includes a table showing the capital intensiveness of the power industry.

167 What will rates be like in 1980? Special report. (In Elect. World, vol. 178, Nov. 15, 1972, pp. 30-32)

Utilities will get an average of 24.4 mils per kw/hr. sold, but even then electricity will still be a bargain. # Tables: 1, Fuel costs (1971-1980); 2, Construction and financing (millions), 1972-1980.

168 Will interest rates top 8 percent in 1972? (In Elect. World, vol. 177, Mar. 15, 1972, pp. 15)

In 1970, net funds raised by governments, businesses, and consumer groups totaled $110 billion; the total rose to an estimated $165 billion in 1971. Presented is what Eastman Dillon Union Securities & Co. expected in 1972.

Management

169 Age and inflation are conspiring against the E. E., by E. Rubinstein. (In IEEE Spectrum, vol. 12, Aug. 1975, pp. 62-66)

Further results from the 1975 IEEE membership survey on salaries.

170 Computer checks do-it-yourself meter reading. (In Elect. World, vol. 179, Apr. 15, 1973, p. 66)

Powell Valley Electric Cooperative uses no meter readers. Instead, customers read their own meters and mail in a card giving the new reading each month. Then a computer program makes use of historical records to run a reasonability check on each customer's power use during the month.

171 The economics of load management by ripple control, by J. Zahari and D. Feiler. (In Energy Econ., vol. 2[1], Jan. 1980, pp. 5-13)

Presents an analysis of the benefits of voluntary load-shedding schemes which interrupt electricity supply when generating capacity is insufficient. # Partial Tables: Estimating the energy cost savings; Estimating the capital cost savings.

172 Electrical World's utility executive conference, 5th. Washington. (In Elect. World, vol. 177, Apr. 15, 1972, pp. 28-29)

Theme of conference was "Planning the system of the Eighties."

173 The electricity discount scheme, 1978: questions and answers. Department of Energy, 1977. 15p.
Contents: Scope and duration of the scheme; The £5 cash payment; The discount on electricity bills over £20--electricity bills for credit consumers; Consumers with coin-in-the-slot (pre-payment) meters; The discount voucher; Consumers on special payment schemes; Special problems.

174 The electricity discount scheme, 1979: questions and answers. London: Department of Energy, 1978. 17p.
Contents: General--scope and duration of the scheme; the £5 electricity payment; The discount on electricity bills over £20; Special problems.

175 For utilities, the future is a systems problem: interview with P. Sporn. (In Elect. World, vol. 175, Jan. 1, 1971, pp. 28-31)
Two of the seven highlights in the interview are: 1, Fuels, in which it is predicted that hundreds of millions of dollars of investment in fossil-fuel facilities, and perhaps as much as $5 billion to supply atomic fuel facilities, will be needed in the next 30 years; and, 2, Cost cutting--Are equipment costs too high? Do more constructive planning and engineering; Are labor costs too high? Make one man-hour of labor do more; Are fuel costs too high? Spend more of your efforts to bring them down.

176 How much do middle management earn? (In Elect. World, vol. 186, Sept. 1, 1976, pp. 53-56)
A management report by the Administrative Management Society. # Tables: Industries and average salary rankings; Annual salary levels for middle management positions in 2,771 US companies.

177 The insurance broker and the electrical industry, by A. Cleaver. (In Elect. Rev., vol. 205[22], Dec. 7, 1979, p. 45)
Looks at the service offered by insurance brokers--concern with major electrical projects will indicate the depth of involvement required if a full service is to be provided.

178-79 Let there be light! But not any more than is necessary, by C. Beatson. (In Engineer, vol. 238[2], June 6, 1974, pp. 34-35)
Argues that cheap electricity is a thing of the past: but industrial lighting is often wasteful despite the wide variety of energy-saving alternatives available. Presents a graph of cost comparison of office lighting systems--comparison by area (at 1973 energy prices).

180 Management of the electric energy business, by E. Vernard. McGraw-Hill, 1979. xi, 403p.

181 Management predictive reporting system, by R. A. Peddie. (In Electronics and Power, vol. 16, Sept. 1970, pp. 340-343)
An account of an information system established by the CEGB's Midlands Region to enable it to predict the overall performance of its plant throughout the year. The system bypasses the conventional hierarchical structure, and eliminates many of the distortions introduced in the usual management reporting chains.

182 Motion means money, by B. Toreslam. (In Elect. Rev., vol. 204[24], June 1979, pp. 28, 30)
One of the most important tasks facing company management is to maximize the return on the capital tied up in the activities of the company. In many cases significant progress can be made by improving the efficiency of material handling.

183 Planning for startup, by R. B. Dean. (In Power Eng., vol. 76, Oct. 1972, pp. 40-42)
Initial start up of a new generating unit requires the coordination of operator training, construction and control system checkouts as well as early preparation of detailed data sheet and schedules--all with built-in flexibility.

184 Profile: the utility marketing executive. (In Elect. World, vol. 177, May 1, 1972, pp. 73-74)
A management broadsheet.

185 Should engineers run utilities: discussion. (In Elect. World, vol. 184, July 1, 1975, pp. 57-60)
A management report.

186 Survival plea: pinpoints the profitable customer and load. Marketing Division of the Southeastern electric exchange meeting. (In Elect. World, vol. 176, Dec. 15, 1971, pp. 85-86)
Discusses aggressive selling linked with promotional activities for utilities--with emphasis on research, analysis, planning, and control of those elements affecting utility loads, revenues and costs. For utilities facing rising unit costs, aggressive across-the-board selling would only contribute to these rising costs, and to a need for eventual rate relief.

187 The probabilistic thinking to improve powerplant reliability, by S. J. Kaplan and B. J. Garrick. (In Power, vol. 124[3], Mar. 1980, pp. 56-61)
Gives factors about purchase and maintenance of power plants equipment.

188 Utilities need a management renaissance, by J. W. Weber. (In Elect. World, vol. 174, Dec. 15, 1970, pp. 87-88; vol. 175, Jan. 1, 1971, pp. 77-78; vol. 175, Jan. 15, 1971, pp. 97-98)

Contents: Pt. 1, Sketch of a dismal financial picture. # Tables: 1, Cost of capital and return as percent of investment; 2, Projections of return on investment, return on equity, and earnings per share decline over the years ahead; 3, Cost trends, 1960-1970; 4, Cost per kw installed of new plants, 1956-68; pt. 2, Illustrates and analyzes how electric utility managements have responded to trends in revenues and costs, in an effort to close the profitability gap; pt. 3, Shows a utility can use and instal a management by objective program.

189 Utilities strive for fuel management. (In Elect. World, vol. 172, Nov. 24, 1969, pp. 32-33)
Contends that computerized analytical 'models' and trained engineers are key objectives as the nuclear utilities scurry to develop in-house fuel cycle management capabilities.

190 Utility supplier cooperation in fuel management: abstract, by J.E. Tribble and W.B. Bigge. (In Combustion, vol. 41, Aug. 1969, pp. 33-34)
Abstract of a paper presented at the American Power Conference, 1969.

191 View from the top: energy and technology will solve dilemma; interview with W.R. Thompson. (In Elect. World, vol. 175, June 15, 1971, pp. 36-38)
An interview with W. Reid Thompson, president and chief executive officer of Pepco, answering questions about construction budget total over three years, main capital investments, and, about internal and external financing of the construction budget.

Pricing

192 Area electricity boards: electricity prices and certain allied charges, by Price Commission. London: H.M.S.O., [1979]. viii, 75p. (1978-79: HC-132)

193 A comparison of electricity prices in the countries of the European Community: five-year survey, 1973-1978. London: Electricity Council, 1978. 20 leaves (Issue no. 7)

194 The effect of price on the residential demand for electricity: a statistical study, by L.S. Mayer and C.E. Horowitz. (In Energy, vol. 4[1], Feb. 1979, pp. 87-99)
Examines the statistics on demand for electricity and the relationship between the price of electricity and demand through analysis of the monthly consumption statistics of a community of owner-occupied town houses.

195 Estimating the price elasticity of US electricity demand, by

V.K. Smith. (In Energy Econ., vol. 2[2], Apr. 1980, pp. 81-85)

The use of a single electricity price has been seen as a major shortcoming of econometric models of residential demand. It has been suggested that demand estimation should be based on the full-scale rate schedule and this suggestion is evaluated by examining demand estimates for 27 investor-owned US utilities over the period 1957-1972.

196 Expected welfare gains from peak-load electricity charges, by E.F. Renshaw. (In Energy Econ., vol. 2[1], Jan. 1980, pp. 37-45)

Examines a two-period case with two types of consumer and one type of generating technology. This illustrates the complexity of analyzing the welfare gains associated with demand meters. # Partial Headings: Single, separate charge for capacity; Total welfare losses due to separate capacity charges; Separate charges vs. equal break-even price; Optimal peak-load pricing; Equity and efficiency in the pricing of electricity.

197 How to crack the conversion market. (In Elect. World, vol. 176, Oct. 1, 1971, pp. 96-97)

Study of conversion and mobile-home markets show differing life-styles and personality traits can affect sales.

198 How to sell all electric HUD projects. (In Elect. World, vol. 177, Jan. 1, 1972, pp. 48-50)

Selling the all-electric concept to Department of Housing and Urban Development (HUD) requires the utility to know and work closely with HUD representatives at regional and area offices serving their service areas.

199 Impact of price expectations on the demand for electrical energy in the United States, by N.D. Uri. (In Applied Energy, vol. 5[2], Apr. 1979, pp. 115-125)

Contends that given the variability of the price of electrical energy over the past several years, the consumer has been put in a position of not knowing precisely from one month to the next what the price of electrical energy will be. Presents a hypothesis of a relationship between the expected price and kilowatt-hour sales.

200 Power plant effluent--thermal pollution or energy at a bargain price? by W.S. Lusby and E.V. Somers. (In Mech. Eng., vol. 94, June 1972, pp. 12-15)

Tables show estimated costs for example energy-center substation for heating and cooling.

201 The price of electricity, by L. Goani. (In Revue de l'Energie, Aug./Sept. 1979, pp. 671+)

202 Price of power. (In Engineering, vol. 213, July/Aug. 1973, pp. 537)

Discusses whether a company could save or lose money by generating its own power.

203 Reselling electricity: a guide for tenants and landlords. London: Department of Energy, 1978. 9p. illus. (RE-2)
Partial Contents: What is the maximum charge a landlord can charge? How do you pay electricity? Are you paying the right price for electricity? Calculating the charges.

204 Sale of firm electric power for resale by private electric utilities, by federal projects, by municipals. Washington, D.C.: Gov't Print. Office, 1974+
Gives wholesale prices and quantities for sale by private utilities.

205 Save it or sell it? by D. Crabbe and B. Lowe. (In Elect. Rev., vol. 205[2], July 1979, pp. 21-22)
Discusses overcapacity in the electricity supply industry.

206 Time of day price elasticities for electrical energy consumption: some empirical findings, by N.D. Uri. (In J. Inst. Fuel, vol. 51[409], Dec. 1978, pp. 199-201)
Consumers are responsive to price changes in peak, intermediate, and base periods and the extent of this response in each period is the same.

207 What price the all-electric office? (In Energy Manager, vol. 2[3], Apr. 3, 1979, pp. 41-42)
Commercial buildings can squander energy to provide conditions that are far from satisfactory. Report on a recent seminar on energy effectiveness in commercial buildings.

Statistics

208 Annual electric industry forecast, 21st. (In Elect. World, vol. 174, Sept. 15, 1970, pp. 35-50)
Contents: Economic indicators foretell new gains; Residential sales lead growth and revenues; Construction expenditures to set new highs; T&D survey shows heavy construction workload. Includes numerous statistical tables.

209 Annual electrical industry forecast, 22nd. (In Elect. World, vol. 176, Sept. 15, 1971, pp. 41-56)

210 Annual electrical industry forecast, 25th by L.M. Olmsted. (In Elect. World, vol. 182, Sept. 15, 1974, pp. 43-58)
Contents: US economy resumes growth--GNP to reach $6.7 trillion in 1995; kw/hr sales growth will be tempered by fuel costs ...; Slower growth in sales and peaks sparks sharp expansion plans and costs. # Partial Tables: Popu-

lation, households, GNP, disposable income, 1963-1995; Industrial uses of electricity, billions of kw/hr, 1963-1995; Industrial sales, 1963-1995; Residential sales, 1965-1995; Total sales, system output, peak load, capability, and margin, 1963-1995; Annual expenditures, millions of dollars, 1963-1995.

211 Annual growth rate on downward trend, by F. Felix. (In Elect. World, vol. 174, July 6, 1970, pp. 30-34)
Analysis of long-term trends in kw/hr growth, GNP, and population shows strong, dependent relationship between economic growth and electric power consumption. Includes useful charts and tables.

212 Changing generation patterns, by H. L. Smith. (In Power Eng., vol. 74, Nov. 1970, pp. 47-51)
Partial Tables: Economic mix of generation capacity-typical system limit on pumped storage sites, 1966-1986; Economic mix of generation capacity-typical system limited pumped storage and intermediate fossil steam plants, 1966-1986.

213 Electrical World annual electrical industry forecast, 27th, 1976. (In Elect. World, vol. 186, Sept. 15, 1976, pp. 43-58)
Contents: Economy shows strong signs of life--GNP reaches for $8 trillion by 1995; Kw/hr sales make sharp recovery--long-range growth rate moves down; Delays call for a surge in nuclear capability at the planning horizon. Includes statistical tables.

214 Electrical World annual statistical report, 1971. (In Elect. World, vol. 175, Mar. 15, 1971, pp. 39-70)
Contents: Economic conditions--1971 a year of moderate recovery; Capital expenditure--US utilities plan 16.6 percent higher spending; East North Central Region holds lead; Sales --residential usage aids 6.4 percent total growth; Capability --generation to gain 37,187 Mw in 1971; Finance--revenues set record in 1970; Canada--capital expenditures to set new records.

215 Electrical World annual statistical report, 1972. (In Elect. World, vol. 177, Mar. 15, 1972, pp. 41-72)
Contents: Business climate will improve moderately in 1972; Budgets of US utilities assured fast growth; Top spending regions swap positions; T&D budgets jump to $6.15 billion; Generating additions set record in 1971; System repairs to cost $1.9 billion; Kilowatt hours sold in 1971 gain 5.3 percent; Revenues top $21 billion, up surplus; Co-op systems add 1/4-million customers; Canadian expenditures set new record. Data supplied by individual utilities are the basis for this report.

216 Electrical World annual statistical report, 1972: capacity additions;

United States and Canada. (In Elect. World, vol. 177, Mar. 15, 1972, pp. 56-61, 70-72)

Contents: Generation additions set record in 1971, aim is 44,529 Mw for 1972; Canadian expenditures set new record but fall short of 1971 budgets. Includes statistical tables.

217 Electrical World annual statistical report: financial. (In Elect. World, vol. 177, Mar. 15, 1972, pp. 65-68)

Customer growth and rate hikes ease the burden of soaring costs. # Tables: Electric rate increase, 1961-1971; Electric rate decrease, 1961-1971; Distribution of electric revenue dollar (percent); Investor owned electric utilities income statement (millions of dollars); Security sales by investor-owned electric utilities (thousands of dollars); Investor-owned electric utilities combined balance sheet (millions of dollars)

218 Electrical World annual statistical report, 1973. (In Elect. World, vol. 179, Mar. 15, 1973, pp. 64-66)

Canadian utilities plan sharp rise in capital spending. Budgets were up a hefty 21 percent over 1972 expenditures. # Tables: Electric plant construction (thousands of dollars); Maintenance (thousands of dollars)

219 Electrical World annual statistical report, 1973: operating statistics. (In Elect. World, vol. 179, Mar. 15, 1973, pp. 58-59)

Vigorous economic gains during 1972 pushed electric energy sales to a staggering 1.57 trillion kw/hrs, nearly 7.4 percent over the previous year's record 1.47 trillion. # Tables: Electric consumers (thousands) 1962-1972; Electric revenue (millions of dollars), 1962-1972; Electric energy sales (millions of kw/hr), 1962-1972; Average annual use and bills, 1962-1972.

220 Electrical World annual statistical report, 1975. (In Elect. World, vol. 183, Mar. 15, 1975, pp. 43-74)

Contents: US economic growth to resume by year-end, surge upward in 1976; Capital expenditure--spending sets new record; Budgets foretell 4.6 percent boost in 1975.

221 Electrical World annual statistical report, 1975: operating statistics. (In Elect. World, vol. 183, Mar. 15, 1975, pp. 66-67)

Industrial sales slip, revenue gains from rate boosts, fuel charges. # Tables: Electric revenue (millions of dollars); Electric energy sales; Average annual use of bills.

222 Electricity: demand vs. capacity. (In Mech. Eng., vol. 97, Mar. 1975, pp. 100-101)

Charts: Summer capabilities, peak loads and gross margins; December capabilities, peak loads and gross margins; Annual kilowatt-hour requirements and load factors.

223 Growth of 4.1 percent in peaks reported by 86 systems. (In Elect. World, vol. 175, Mar. 15, 1971, pp. 112-114)
Tables: West region leads growth in Winter peak, with Northeast second; Canadian demand grows 8.7 percent with six systems up over 10 percent.

224 Long-run electricity of US energy demand: a process analysis approach, by S.C. Parikh and M.H. Rothkopf. (In Energy Econ., vol. 2[1], Jan. 1980, pp. 31-36)
Argues that the process analysis approach, which makes use of subjective judgment, has certain advantages over more formal econometric approaches and is of relevance to current work in energy demand forecasting.

225 New capacity: a profile of utility growth, by R. Schuster. (In Power Eng., vol. 78, Apr. 1974, pp. 40-45)
Tables: Committed generating plants, 1974-1988; 2, Ten leading utilities in total Mw of committed new generating capacity; 3, Types of fuel for committed fossil plants; 4, Leading engineers and/or constructors by total Mw of committed capacity--fossil, and nuclear combined, and by percent of total plants; 5, Leading engineers and/or constructors by number of plants--fossil and nuclear ...; 6, Leading boiler and reactor suppliers....

226 Peak grow 6.6 percent for 91 utilities in survey. (In Elect. World, vol. 176, Nov. 1, 1971, pp. 68-70)
Increase in noncoincident summer peak shows faster growth rate for first time in three years, despite reports from all regions of unseasonable weather and industrial decline. # Tables: Systems in Pacific Southwest region report 9.9 percent gain in summer peaks; Ten systems in Canada report 5.6 percent increase in summer peak loads; June is peak month for over one-third of 91 systems.

227 Power generation growth patterns, by R.C. Rittenhouse. (In Power Eng., vol. 79, Apr. 1975, pp. 42-48)
Several negative factors are "whipsawing" utilities in their effort to meet growing energy demands. # Partial Tables: Committed generating plants, 1975-1994; Ten leading utilities in total MW of committed new generating capacity; Types of fuel for committed fossil plants.

228 Quick-start and cyclic capacity for the 70's, by F.M. Swengel. (In Power Eng., vol. 75, June 1971, pp. 34-40)
Table: Historical and projected capital cost of generation, 1960-1972.

229 Survey of 84 utilities shows 64 percent gain in peaks. (In Elect. World, vol. 174, Oct. 15, 1970, pp. 82-84)
Increase in noncoincident summer peaks trails previous rise for second year in a row. Without load curtailment

over peaks the total increase of 14,130 Mw could have been greater. Includes tables.

230 T&D spending gains slightly. (In Elect. World, vol. 183, Mar. 15, 1975, pp. 53-57)
Tables: New expenditures for lines and substations, 1974-75; New lines and substations energized, 1974/75.

231 Underground distribution spending will rise 20 percent. (In Elect. World, vol. 173, Feb. 2, 1970, pp. 57-61)
Tables: New overhead transmission and distribution spending--millions of dollars; New underground transmission and distribution spending--millions of dollars; New substation spending--millions of dollars.

Taxation

232 Investment tax credit gets a boost. (In Elect. World, vol. 183, May 1, 1975, pp. 25-26)
Discusses the temporary investment tax credit of 10 percent for utilities.

233 Taxes could shape energy policy. (In Elect. World, vol. 182, Nov. 1, 1974, pp. 28)
Contends that a combination of pollution taxes and direct pollution regulation would be more meaningful and effective than regulation alone.

234 Utilities want tax breaks. (In Elect. World, vol. 175, Mar. 1971, pp. 27-28)
The Treasury Department's attempt to give utilities a tax break that could amount to $1 billion a year, by including them in new liberalized depreciation rules.

III. ENERGY (General)

Costs

235 Arctic islands potential should justify high cost, by F.G. Rayer. (In World Oil, vol. 188[5], Apr. 1979, pp. 153-158)
Geophysical and drilling evidence supports the contention that Canada's Arctic Islands are in a major hydrocarbon province.

236 Assessing environmental costs of energy procurement, by R.G. Alderfer. (In 4th Conf. on Energy, Univ. of Mo., Rolla, 1978, pp. 79-84)
Describes an ecological approach to the assessment of environmental impact and associated costs of all major phases of energy procurement.

237 Assessing the costs and benefits arising out of conservation, by J. Harris. (In Engineer, vol. 240, May 15, 1975, pp. 32-33, 77)
Discusses the report by the National Economic Development Office (NEDO) on energy which presents ideas for industry on reducing consumption.

238 Burner operation in the process industries, by G.H. Leaver. Urquhart Engineering Company, [1978?] 14p.
The vast increase in energy costs which has taken place over recent years makes it essential to make a re-appraisal of plant profitability.

239 Campus utility costs, by B.O. Turner. (In Heating-Piping, vol. 45, Nov. 1973, pp. 85-93)
Studies at three Utah institutions of higher learning yield comprehensive data that can be used to estimate utility consumption and costs for existing buildings not metered, buildings under construction, and future buildings.... Tables: 1, Heating: annual energy consumption, costs and peak demand comparisons; 2, Cooling: annual energy consumption and cost consumptions; 3, Domestic water: annual consumption, cost and peak demand comparisons; 4, Total electricity: annual consumption, cost and peak demand comparisons; 5, Electricity for mechanical and for lighting: annual consumption and cost comparison.

240 Combustion control is cost control, by A. Bond. (In Process Eng., Apr. 1975, pp. 74-75)
Suggests that after taking the "energy sense is common sense" precautions around the plant, the next item on the fuel-economy checklist may well be the combustion process itself.

241 Commercial aviation and energy. Geneva: International Air Transport Association, 1975. [23] leaves.
Contents: Sect. 1, Summary and conclusions; sect. 2, Commercial aviation and energy; sect. 3, Appendices; sect. 4, Tables--Aviation fuel prices, 1934-1973; IATA global average fuel price index; Impact of rising fuel cost on total world international scheduled airlines; Impact of rising fuel cost on total IATA international scheduled operations; Impact of rising fuel costs on North Atlantic (passenger) traffic; Impact of rising costs on local European (passenger) traffic; Impact of rising fuel costs on international scheduled services summary.

242 Communications on energy: energy costs of house construction --a reply. (In Energy Policy, vol. 5[1], Mar. 1977, pp. 76)
Original paper by E.M. Gartner and M.A. Smith was published in Energy Policy in June 1976. Reply by M. Slesser and T. Markus.

243 Comparative cost study of four wet/dry cooling concepts that used ammonia as the intermediate heat exchange fluid, by R.D. Tokarz, D.J. Braun and B.M. Johnson. Richland, Wash.: Battelle Pacific Northwest Labs., 1978. 151p.

244 Compare your home heating costs: an independent guide to the costs of different fuels and heating methods and how to reduce them, by the Department of Energy. London: HMSO, 1978, 18p.
Partial Contents: Typical cost of heating your main living room if you haven't got central heating; Typical cost of providing and running a separate hot water supply; Considering whole house heating and live in a two-bedroom terrace house; Considering whole house heating and live in a post-war three-bedroom semi-detached house.

245 Compare your home heating costs: an independent guide to the costs of different fuels and heating methods and how to reduce them. London: Department of Energy, 1979, 15p.
A pamphlet designed to promote energy saving.

246 Comparing process energy costs: data sheet. (In Power, vol. 118, Sept. 1974, pp. 24)
Presents a chart to aid comparison, and typical heating values.

247 Computer flight plans for business aviation, by R. Parrish. (In Business and Comm. Av., Mar. 1979, pp. 54-58)
Questions whether they save energy, cost, and time or are merely a convenience.

248 Conference on the North and Celtic seas. Organised by the Financial Times, Petroleum Times and British Airways. [1974]. 1v. (variously paged)
Twenty-one papers covering various aspects of costs which include: Exploration and development costs; Equipment estimated capital costs based on estimated total costs based on estimated total number of units operational or installed by 1979 in UK offshore waters; Financing in the North Sea--a view of a Norwegian Banker, by Nils B. Gulnes; Financing in the North Sea--the views of an American Banker, by Jack A. Horner.

249 Conservation of energy: special report. (In Chem. Eng. Proc., vol. 71, Oct. 1975, pp. 69-73)
Discusses, among other topics, the actual energy costs. Table 3, Cost in ¢ to pump 1 pound of air to atmospheric pressure.

250 Cost comparison of energy projects: discounted cash flow and revenue requirement methods, by D. L. Phung. Oak Ridge, Tenn.: Oak Ridge Associates Univ., 1980, 49p. (ORAU-IEA-80-8[M])
Provides simple formulations for the two methods: discounted cash flow and revenue requirement, and some special cases of interest to costing practices.

251 Cost of energy over the next decade, by N. A. White. (In Energy World, no. 42, Nov. 1977, pp. 2-9)
Identifies and discusses the principal determinants of energy costs from now until the late 1980s. Factors discussed include level and pattern of world energy consumption; conventional energy reserves and costs. Concludes that the cost of all forms of energy over the period will be largely dictated by the pricing policies of the Organization of Petroleum Exporting Countries.

252 Cost per barrel approach to evaluation of energy conservation options, by J.M. Power and G.S. Gill. (In Energy & the Environ., Proc. of the Nat. Conf., 4th. Cincinnati, Ohio, 1976, pp. 91-97)
Presents a standard method of investment evaluation and then converts it into a figurative criterion to which a business firm or individual investor can easily relate.

253 Costing methodologies for energy systems, by J. Allentuck and others. Upton, N. Y.: National Center for Analysis of Energy Systems, 1979, 23p.

Discusses the problem of devising a methodology for arriving at costs of such systems which may be used to compare alternative sources.

254 Creating a uniform energy demand to cut electricity costs, by C. Beatson. (In Engineer, vol. 240, Jan. 23, 1975, pp. 32-33)
Heat storage is one of the ideas put forward to smooth out load curves and reduce pressure on generation at peak times. Charts: Typical demand fluctuations on CEGB system, winter 1972/73; Order of merit in terms of fuel costs.

255 Cut slasher energy costs, by D. L. Nehrenberg. (In Textile Ind., vol. 141, Jan. 1977, pp. 106, 108, 110, 112)
Significant savings in energy consumption at the slasher can be made by changing from hot air drying, going to single-dip from double dip, and reducing wet pick to 100 percent from 130-140 percent.

256 Design and calculated performance and cost of the ECAS phase II open cycle MHD power generation system, by L. P. Harris. (In Combustion, vol. 50[9], Mar. 1979, pp. 21-30)
A 2,000-MWe MHD steam plant for central station application designed and costed as part of the energy conservation alternatives study (ECAS). It yields an estimated overall efficiency of 0.483, a capital cost of $718 per kWe (1975 dollars) and a cost of electricity at 65 percent capacity factor of 32 mills per kWe-hr. # Partial Tables: Major components and subsystem capital costs for open-cycle MHD; Plant capital cost estimate summary open-cycle MHD: Summary performance and cost open-cycle MHD.

257 Design for high cost energy: the Scandinavian approach, by R. Asantila. (In Tappi, vol. 57, Oct. 1974, pp. 117-121)
In Scandinavia, and especially in Finland, certain practices have been used effectively to counteract the high cost of energy. The higher the cost of energy, the more attention has to be paid to (a) process design vs. heat and power balances; (b) cost and availability of external energy; (c) recovery of process fuels; and (d) combined in-plant heat and power generation.

258 Design plants to save energy, by J. E. Hayden and W. H. Leavers. (In Hydrocarbon Process, vol. 52, July 1973, pp. 72-75)
Deliberate planning for conservation of energy can cut costs by a fifth or more in operation and investment.

259 Development of a joint-cost allocation manual for integrated community energy systems: phase II. Washington, D. C.: Ernst and Whinney, 1980, 23p.
The advantages and disadvantages of alternative justifiable

expenditure are discussed and a framework is developed for application of the method to the thermal and electric services of ICES plant.

260 Energy, by T.M. Sugden and F.E.J. Briffa. (In Chem. Ind., Aug. 16, 1975, pp. 669-676)
Reviews briefly how the cost of oil escalated. # Partial Topics: The escalation in price; Alternative energy. # Partial Charts: The rising price of oil; Economic growth, energy and oil.

261 Energy alternatives: a comparative analysis. Washington, D.C.: Department of the Interior, Bureau of Land Management. 1973+
Annual. Gives data on energy resources systems in the United States, and costs.

262 Energy analysis for high-rise apartment buildings, by H. Kunstadt and P. Herzog. (In Bldg. Systems Design, vol. 67, Dec. 1970, pp. 17-28)
Includes cost considerations.

263 Energy and cost analysis of residential heating systems. Springfield, Va.: National Technical Information Service, 1978. 59p.
Estimates cost effectiveness ... in residential space heating systems.

264 Energy at the crossroads: where do we go from here? by L.G. Hauser. (In Lighting Design & Application, vol. 4, Jan. 1974, pp. 6-11)
Contends that there is an alternative to former reliance on imports of oil and natural gas to meet energy needs.

265 Energy conservation and life-cycle costing methods, by J.A. Belding. (In Energy Int. J., vol. 3[4], Aug. 1978, pp. 421-426)
A qualitative summary of life-cycle procedures, with particular reference to the role of the Department of Energy in assuring proper utilization of this type of analysis.

266 Energy consumption in vehicle manufacture, by B.N.A. Lamborn. (In Am. Soc. CE Proc., vol. 101, [TE 2(11278)], May 1975, pp. 409-413)
Includes table showing fuel costs for intercity transportation.

267 The energy cost of goods and services, by R.A. Herendeen. Oak Ridge, Tenn.: National Service Foundation, Oak Ridge National Laboratory, 1973. 1v.
Calculates total cost of many consumer goods and services. Includes statistics of primary energy use by consuming sectors, energy allocation and pricing.

268 Energy: end of an era, by F.H. Happold. (In Chem. Ind., Jan. 5, 1974, pp. 6-7)
Contends that the long-term problems in energy stem from the sharp rise in the cost of oil rather than its relative scarcity. On present form the Arab states could soon be running a favorable trade balance of some $20 billion to $30 billion a year, with corresponding deficits spread among the oil importing countries.

269 Energy forests and fuel plantations, by G. C. Szego and C. C. Kemp. (In Chem Tech., vol. 3, May 1973, pp. 275-284; Discussion, vol. 3, 1973, pp. 391-392; Reply, vol. 3, July 1973, pp. 392+)
Presents a convincing argument against nuclear power. # Partial Topics: Fuel production costs and prices--fuel cost of pulpwood, fuel cost of corn, fuel cost from perennials; Price competitiveness. # Partial Charts: Estimated prices for fuel as a function of crop growth cycle and solar energy conversion efficiency (cost and profit funded regularly); Estimated prices for fuel as function of crop growth cycle and solar energy conversion efficiency (cost paid from working capital and profit deferred); Cost of fuel burned for electrical generation in the United States. # Partial Tables: Estimated capital cost of harvesting solar energy; Pulpwood prices in southeastern states in 1971; Estimates of the price of fuel derived from pulpwood and chips at 1971 prices; Estimated representative price for fuel value derived from corn; Crop growing cost assumptions ...; Estimated costs and equipment sales prices for energy plantation-produced fuel (solar energy conversion to fuel value: 0.4 percent).

270 Energy in perspective, by R. Shinnar. (In Chem. Tech., vol. 5, Apr. 1975, pp. 225-231)
Discusses the energy crisis with special reference to the rising cost of oil imports. # Partial Topics: Reducing future demand for imports; Increase domestic production now. # Partial Charts: Supply/Demand B/D oil equivalent vs. years, 1960-1985; Heavy fuel demand, 1970-1980; Energy consumption in USA, Western Europe and Japan, 1955 to 1985. # Tables: 1, Magnitude of energy crisis; 2, Cost of new energy sources ($ per barrel of oil equivalent); 3, Investment cost of pollution-free fuel from coal (per barrel of oil consumed per day--mining of coal and power production not included); 4, An energy balance for 1977; 5, Power industry.

271 Energy: national assets, corporate liabilities, by R.J. Sherman. (In Combustion, vol. 48, Jan. 1977, pp. 6-12)
Charts: 1, Energy sources for electric utility power generation; 2, Cost of fuel for electric generation--total electric utility industry; 3, Cost of fuel by regions--total electric utility industry; 4, Expenditures for pollution abatement--air and water. # Tables: Anticipated coal prices, 1976; Cost saving summary--mill energy conservation study.

272 Energy output and cost given in two computer programs. (In Elec. World, vol. 175, Mar. 15, 1971, pp. 114)
Discusses the energy output and production cost at individual generating stations which were developed successfully in two simulations by computer.

273 Energy savings on the home front. (In Elect. Rev., vol. 204 [24], June 1979, pp. 19-20)
Contends that most electrical appliances will soon have to be labeled with estimated annual operating costs.

274 Energy solutions linked to $ supply, by R.H. Wills. (In Energy Pipelines & Systems, vol. 1, June 1974, pp. 29-30)

275 Energy systems review, by H.H. Phipps. (In ASHRAE J., vol. 12, July 1970, pp. 49-54; Discussion, by R.E. Roushey, vol. 12, Sept. 1970, pp. 4-5)
Contends that there is considerable evidence among business, industry and government agencies that emphasis is no longer primarily or exclusively placed on first cost. There is apparently an increasing trend toward consideration of long-term owning and operating costs as opposed to the former almost total concern with first costs.

276 Energy use and other comparisons between diesel and gasoline trucks, by K.M. Jacobs. Springfield, Va.: National Technical Information Service, 1977. 107p.
Presented are comparisons of fuel consumption and operating costs for gasoline and diesel-powered trunks.

277 Evaluating energy savings measures: impact of rising energy costs, by D.C. Lehmann. (In ASHRAE J., vol. 18, June 1976, pp. 46-47)
Capital projects for energy reduction are very difficult to justify with conventional yardsticks because these methods give distorted results. A different procedure must be used to justify energy conservation methods.

278 Evaluation of resource impact factors versus social cost estimates in determining building energy performance standard levels, by L.A. Nieves and others. Richland, Wash.: Battelle Pacific Northwest Laboratories, 1979. 123p. (PNL-3087: EY-76-C-06-1830)
Report explores the feasibility of developing a factor that could be used to adjust a design energy budget to account for the external costs associated with energy consumption.

279 Expected energy production costs by the method of movements, by N.S. Rau and others. (In IEEE Power Eng. Soc. Prepr. Summer Meet., 1979. 8p.
Outlines a new method of calculating the expected energy generation in a generating system by probabilistic simulation.

280 Fighting high energy costs with centrifugal clutches, by E.C. Goodling. (In Machine Design, vol. 46, Sept. 19, 1974, pp. 119-124)
Explains how to maximize savings.

281 Fluid heaters and rising energy costs, by C. Bonnet. (In Hydrocarbon Process, vol. 54, May 1975, pp. 34D, 34EE, 34G, 34GG, 34I, 34II)
Fired heater efficiencies can be improved to cope with ever-increasing fuel costs. Presents facts for consideration.

282 Future energy market environment, by L.W. Fish. (In Am. Gas. Ass. M., vol. 52, Feb. 1970, pp. 9, 31)
Article based on a presentation made by the author at the annual meeting of the Institute of Gas Technology in Chicago, Nov. 20, 1969. Attempts to provide some realistic answers to such vital questions as the profit future for the gas business, in the impending squeeze between reducing supply and increasing the cost thereof and the more demanding market.

283 Future shock: energy cost projection, by V. Traudt. (In Heating/Piping, vol. 48, May 1976, pp. 67-70)
Discusses the harsh realities of rising fuel costs which intruded upon the University of Nebraska's physical plant budget, and what personnel did about it. # Tables: 1, Fuel oil consumption and costs; 2, Natural gas consumption and costs; 3, Consolidated energy consumption from tables 1 and 2 in equivalent oil; 4, Savings each year due to 1972/73 energy conservation adjusted for changes in degree-days and floor space.

284 Guidelines in cost control--2: optimum goals needed to reduce expenses, by C.F. Dwyer. (In Pet. Eng., vol. 44, Sept. 1972, pp. 74, 78, 80)
Tables: 8, Analysis of sub-surface costs; 9, Operating cost breakdown; 10, Guiding standards--surface maintenance costs, surface operating, subsurface costs; 11 & 12, Surface costs, 1963.

285 Guidelines in cost control--3, by C.F. Dwyer. (In Pet. Eng., vol. 44, Oct. 1972, pp. 62, 68, 74, 76)
Charts: Western Division expenditure control, including plants, 1965; North district normal operations progress report, 1965. # Tables: 13, Determination of cost yardsticks; 14, Example of use of guides; 15, Field FG operating costs vs. guides; 16, Comparison of cost guides for Western and Northern Divisions.

286 Here's how K-factor calculations help you make energy comparisons, by J.W. Howland. (In Coal Age, vol. 81, June 1976, pp. 167, 171)
A common yardstick used to compare the cost of coal

to natural gas or crude oil....# Tables: 1, Calculations of energy yardsticks; 2, Energy K-factors for coal; 3, Energy K-factors for crude oil; 4, Comparative cost of fuels.

287 How to figure a fair charge for purchased heating, cooling, by J.H. Beck. (In Heating/Piping, vol. 42, Dec. 1970, pp. 64-67)
When a tenant buys heating and cooling from a central or district plant, not all of his operating and maintenance costs are foregone; a portion remains. Here's one approach to estimating fuel and residual costs of operation and maintenance.# Topics: What to evaluate; The fixed costs; The variable costs; A fair and equitable charge.# Chart: Monograph for determining annual fuel consumption for heating.# Tables: 1, Summary of fixed costs; 2, Summary of variable costs.

288 Input-output techniques and energy cost of commodities. (In Energy Policy, vol. 6[2], June 1978, pp. 162-165)
Quantifies the desirability of several methods of calculating energy cost of commodities from input-output data, using an a priori technique based on exact treatment of an artificially homogenized A-matrix.

289 Kimberly High revisited: operating costs for six years, by W.R. Ratai. (In Heating/Piping, vol. 41, Dec. 1969, pp. 86-88)
A review which reveals that the heat reclaim system which cools as well as heats has operated successfully. The operating costs, as anticipated, were lower than the cost of fuel for just heating a conventional school.# Table: Six-year operating cost....

290 Life-cycle costing for consumers of energy-conserving devices, by S.S. Penner and M.R. Brambley. (In Energy Int. J., vol. 3[4], Aug. 1978, pp. 415-419)
Gives general relations for life-cycle costs and savings associated with energy conserving devices for consumers.

291 Life cycle costing, government policies and the diffusion of energy-conserving technology, by M.O. Stern. (In Energy Int. J., vol. 3[2], Apr. 1978, pp. 173-202)
Considers some economic tools that permit estimation of the costs and benefits of government actions. Costs may consist of funds spent on research, development and public education, of tax revenues foregone.# Table: Some promising candidate innovations for life-cycle cost and net energy-cost analysis.

292 Life cycle costing of energy systems, by W.R. Baker. (In Proc. of the Annual UMR-DNR, Univ. of Mo., Rolla, 1978, pp. 108-114)

Life-cycle costing process is applied to the evaluation of capital expenditures.

293 A low-cost energy-saving window system, by T. Morrone. (In Energy, vol. 5[3], Mar. 1980, pp. 207-215)
A low-cost, highly insulated window system, which can be fabricated with easily available material, is described.

294 Medium BTU gas fits refiner's needs, by T.M. Barnett, and others. (In Hydrocarbon Process, vol. 57[6], June 6, 1978, pp. 131-133)
Discusses economics and technological aspects of process heating in petroleum refineries and petrochemical plants. Cost is estimated to be high.

295 The new market for energy services, by R.W. Sant. (In Energy, vol. 5[3], Summer 1980, pp. 19-20)
Contends that the traditional retailers of oil, gas, coal and electricity will gradually become energy wholesalers because of organizational and recruiting problems.# Tables: Energy services provided as a percent total national fuel cost in 1978.

296 Optimized heating-cooling control to improve comfort and economy, by D. Spethmann. (In Heating/Piping, vol. 42, Nov. 1970, pp. 89-90)
Stipulates that consulting engineers, ever concerned with total system performance and initial costs, are taking a harder look at on-going operating costs. A low-cost system that is expensive to operate will soon offset any initial savings.

297 Photovoltaic power: despite great promise, it will make no significant impact on the energy supply situation in the US until system cost drops by a factor of 100 or more--to $15-45/m^2 for higher efficiency arrays. (In Astron & Astronaut., vol. 13, Nov. 1975, pp. 28-32)
Contends that the major remaining obstacle to immediate large-scale use of this approach, the capital cost of the system, centers on cost reduction. Chart shows cost projections for silicon solar cells.

298 Pollution control energy costs, by E. Hirst. (In Mech. Eng., vol. 96, Sept. 1974, pp. 28-35)
Discusses what would be the energy cost of a program to clean up and maintain the quality of the environment.

299 Possibilities and problems for utilizing peat as an energy source, by L. Ojefors. (In Energy, vol. 5[3], Summer 1980, pp. 6-7)
Tables: 1, World supply of peat; 2, Composition of total peat production cost; 3, Heat production costs.

300 Practical aspects of energy source analysis, by R.H. Braun. (In Bldg. Syst. Design, vol. 69, May 1972, pp. 33-35)

301 Processors seek ways to hold down energy costs. (In Eng. & Min. J., vol. 177, June 1976, pp. 247-248)
Most pyrometallurgical plants are closely tied to conventional fuels. Fuel shortage and higher costs for natural gas and fuel oil have focused attention on alternative energy sources.

302 Review paper: Canadian renewable energy prospects, by H. Swain, R. Overend and T.A. Ledwell. (In Solar Energy, vol. 23[6], 1979, pp. 459-470)
Partial Headings: A digression on costs; Solar energy: space and water heating.# Tables: Revised oil demand and availability high price scenario, 1970-1990; A speculative view of the place of solar energy (all forms) in the global energy budget; Possible solar sales for space and water heating in Canada; Marginal costs of electricity ($ 1976)

303 Saving energy and dollars, by J.F. Finn. (In Lighting Design & Application, vol. 4, Jan. 1974, pp. 30-32)
Examines other tradeoffs that should be considered in evaluating energy and monetary savings. Also, shows how energy and costs vary for four different sources in a lighting system for a turbine room in a typical power plant.# Tables: 1, Lighting's share of the energy load; 2, Total cost comparison of the four sources; 3, Unit cost comparison for the four sources; 4, Estimated potential energy savings.

304 Soaring cost of energy projects. (In Petroleum Economist, vol. XLVII[3], Mar. 1980, pp. 94-95)
Contends that faced with steeply rising energy prices, the need for far larger investments in increasingly expansive energy projects must be accepted. Success in this objective might at least prevent energy prices from going through the roof.

305 Some influences on energy costs by anti-air pollution investment, by M. Takei. (In World Energy Conf., 9th Trans. Paper & Discuss., Sept. 1974, pp. 761-778)
Data presented in tabular and graphical form.

306 State energy costs vary enough to affect economic development, by S.P. Burggraf. (In Prof. Eng., vol. 48[3], Mar. 1978, pp. 23-28)
Data on prices paid by individual utilities for generating fuels indicate that energy costs vary widely by state and, to a lesser extent, by broad geographical region.

307 Steam station cost survey. 16th special report. (In Elec. World, vol. 172, Nov. 3, 1963, pp. 41-56)

Topics: Total power cost turns up; Fixed charges upturn in total energy cost; Production costs continue downward trend; Investment declines but heat rate suffers.# Includes extensive statistical tables showing figures under various headings such as Fuel consumption, Investment, Energy costs.

308 Steam station cost survey, 18th special report. (In Elect. World, vol. 180, Nov. 1, 1973, pp. 39-54)
Contents: Nuclear units slow power cost rise; Fuel price rises boost cost of busbar energy; Operating costs jump 20 percent in two years; Fixed charges reflect higher contribution cost.

309 Steam station cost survey, 19th special report, by M.O. Olmsted. (In Elec. World, vol. 184, Nov. 15, 1975, pp. 43-58)
Contents: Busbar energy costs set record; Fixed charges, fuel share blame for record energy costs; Operating costs jump 103 percent in two years; Construction costs climb 28 percent in two-year interval.

310 TERA and house heating costs, by D.O. Dawson. (In Am. Assn. M., vol. 57, June 1975, pp. 23-25, 27)
Total Energy Resource Analysis system (TERA)--a planning tool that can respond to the broad variety of "what if" questions--is an automated energy model designed to assist the American Gas Association and the gas industry in evaluating policy issues relative to factors such as gas supply and demand and energy prices.# Partial Tables: Oil/Gas Btu equivalent prices at sources, 1975-1985; National average annual house heating fuel bills--for gas, oil and electric.

311 Thermochemical energy transport costs for a distributed solar power plant, by O.M. Williams. (In Solar Energy, vol. 20[4], 1978, pp. 333-342)
Thermochemical energy transport costs are calculated for a solar thermal power plant based on a distributed network. The derived costs are dominated by the pipe installation components.

312 Three modes of energy cost analysis: then-current, base-year, and perpetual-constant dollar, by R.M. Harnett and D.L. Phung. (In Energy Systems and Policy, vol. 3[1], 1979, pp. 61+)

313 Total energy cost of household consumption in Norway, 1973, by R. Herendeen. (In Energy Int. J., vol. 3[5], Aug. 1978, pp. 615-630)
The economic data of the 1973 Norwegian survey of consumer expenditures are converted to their corresponding energy requirements. In this study, household consumption of all goods and services were energy-costed to obtain the energy cost of living.

314 Total energy firms stop exhaust heat just puffing away, by C. Beatson. (In Engineer, vol. 233, Sept. 23, 1971, pp. 24-27)
Are power and energy costs a significant part of total outgoings? Fresh incentives are provided by rising costs of fuel and electricity and by the new catchphrase, "total energy"--defined as planned integration of on-site power generation and process heating systems to minimize waste. Graphs show the effect of changes in price of electricity and fuel oil if gas price stays constant. There is also a diagram of total energy system.

315 Transition in energy resource, by F. Stock. (In Petroleum Economist, vol. XLVII[5], May 1980, pp. 207-208)
Considers rising costs.# Tables: World energy demand, 1965-2000; World energy demand, 1928-1978.

316 US inter-industry cost structure of the energy sectors, 1958, 1967 and 1968, by K.O. Kymn and W.P. Paage. (In Energy, vol. 4[3], June 1979, pp. 451-456)
Examines the similarity in the cost structure of energy sector in the 86 order input-output tables (1967) and also, changes for the period 1958-67 and 1967-68 in energy sector cost structures.

317 Wasting energy on central heating? by R. Livesey. (In Engineering, vol. 208, Oct. 31, 1969, pp. 462-463)
Some suppliers of housing insulation materials imply that fantastic savings can be made on the fuel bill if their products are used. But how good are they? Is double glazing really worthwhile?

318 Wood and energy: [power from waste]. (In Energy Development, vol. 3, Oct. 1979, pp. 18-20)
A woodworking factory in Colombia is saving 100,000 US dollars a year by no longer producing energy in a diesel power plant. The fuel is wood.

319 The world distribution of commercial energy consumption, by M.C. Jackhart, M. Arditi and I.A. Arditi. (In Energy Power, vol. 7[3], Sept. 1979, pp. 199-207)
Partial Charts: Per capita energy consumption for countries of rank 10, 50, 100 and 150 in 1950, 1969 and 1975; Number of countries in each per capita energy consumption range in 1950; Number of countries in each per capita energy consumption range in 1969; Number of countries in each per capita energy consumption range in 1975.# Table: Per capita energy consumption and rank for selected countries, 1950, 1969 and 1975.

Economics

320 Alternative technologies for US residential water heating:

energy savings and economic benefits, by E. Hirst, J. Carney and D. O'Neal. (In Energy Policy, vol. 7[4], Dec. 1979, pp. 307-320)

Partial Charts: 3, Electricity use vs. capital cost for conventional heat pump, and solar water heaters for regions 1-5. Baseline system electricity use and capital cost for conventional, heat pump, and solar water heaters for regions 6-10.

321 Civil aviation: energy considerations, Paper 9, by the Department of Energy, Advisory Council on Energy Conservation. London: HMSO, 1979, vi, 26p. (Energy Paper, 36)

Partial Contents: Oil prices and future oil consumption by civil aviation; Taxation of aircraft fuels.# Appendices: UK airlines--statistics of traffic and aircraft operations; Taxation of transport fuels and vehicles, other than aircraft fuels; Sources of statistical data.

322 A comparison of energy projections to 1985, by J.R. Broadman and R.E. Hamilton. Paris: Organisation of Economic Co-operation and Development, 1979, 17p. (International Energy Agency--Monographs, 1)

323 Development of a joint-cost allocation manual for integrated community systems (ICES). Washington: Ernst and Ernst, 1978. 117p.

Includes information on cost and econometrics.

324 District heating and economic analysis: the French case, by P. Ramain. (In Energy Econ., vol. 2[3], July 1980, pp. 154-160)

Examines the role of economic analysis in making decisions about district heating.# Partial Headings: Evaluation methods--Costing heat from Combined Heat and Power (CHP); Official costing methods.

325 Dynamic input-output analysis of the economy of energy, by R.P. Rhoten. (In Energy, 78, IEEE, 1978, pp. 40-42)

Contends that an economic analysis of any energy price increase must include not only the direct effect on the consumer, but the direct effects resulting from energy use as a factor of production. If natural gas prices increase, home heating costs will obviously rise.

326 Economic impact of energy shortages on commercial air transportation and aviation manufacture. Springfield, Va.: National Technical Information Service, 1975, 1v.

Gives an analysis of commercial air carriers' future prospects to 1980/90 in energy shortage situation, in terms of direct and indirect effects of energy shortage and increased fuel costs.

327 Economic policy and exhaustible resources, by H. Motamen. (In OPEC Review, vol. 3[1], Mar. 1979, pp. 45+)

328 Economic restraints on US energy supply and demand, by D. R. Knop and J. F. Roorda. (In J. Pet. Tech., vol. 27, July 1975, pp. 803-812)
Partial Topics: Economic projections; Energy price assumptions; Projects of natural gas hydropower and nuclear power availability; Outlook for energy demand; Projections of oil and coal availability; Economic consequences of higher energy prices.# Partial Tables: 1, Selected price elasticity of demand in the DRI energy policy model; 2, Initial projections of key economic variables exogenous to the DRI energy policy model (Billions of dollars); 3, Projected energy prices.

329 The economics of energy, by M. G. Webb and M. J. Ricketts. New York: Macmillan, 1980, xiv, 315p.

330 Economics of geothermal energy, by G. E. Morris, J. W. Tester and G. A. Graves. Los Angeles, Cal.: American Nuclear Society, 1980, 12p. (CONF-800216-1)
Presents a summary of the resource, technical and financial considerations which influence the economics of geothermal energy in the US estimates of resource base and levelized busbar cost of base load power for several types of geothermal resources.

331 Economics of insulating glass, by R. T. Tamblyn. (In ASHRAE J., vol. 15, June 1973, pp. 41-45)
Presents a simplified method for estimating the feasibility of using insulating glass for a given insulation. Graphs show potential savings in heating, cooling, air handling, energy utilization and operation.

332 The efficient use of energy drying operations, by H. B. Weston. London: National Industrial Fuel Efficiency Service, [1979?]. 12p.
Paper prepared for the energy manager; highlights the more important aspects of drying and drier operation with particular reference to energy economy.

333 Energy and the economy: an interrelated perspective, by J. H. Krenz. (In Energy, the Int. J., vol. 2[2], June 1977, pp. 115-130)
A reduction in energy intensiveness can be achieved through energy-related capital investment and social changes.

334 Energy, economic growth, and equity in the United States, by N. P. Kannan. New York, London: Praeger, 1979. 218p.

335 Energy economics, by L. M. Liberman. (In Am. Gas Assn. M., vol. 58, June 1976, pp. 4-7)
Tables include price per barrel of oil equivalent.

336 Energy economics and policy, by J. M. Griffin and H. B. Steele. London: Academic Press, 1980. xiv, 370p.

337 Energy: income coefficients and ratios: their use and abuse, by M.A. Adelman. (In Energy Econ., vol. 2[1], Jan. 1980, pp. 2-4; Rejoinder, vol. 2[3], July 1980, pp. 184-185)
Considers two commonly used relations between income and energy consumption: the ratio of consumption to income, and the incremental energy-income coefficient.

338 Energy: latent economics, the use of solar energy in Britain, by R. Tomkins and M.G.J. Wilson. (In Engineering, vol. 216, June 1976, pp. 405-409)
Rapidly rising costs coupled with increasing awareness that the earth's fuel resources are not inexhaustible, have reawakened interest in renewable sources of power. Now, with the increased costs of what are at present considered to be conventional fuels, warnings that in any event stocks are not limitless, and reappraisal of the whole scene are underway.# Partial Tables: 1, Present value of £1 saved annually; 2, Returns on investment in solar water heaters; 3, Range of possible return on solar water heaters.

339 The energy needs of transport, by I.C. Cheeseman. (In Transport, Mar/Apr. 1980, pp. 9, 11, 13)
Topics: Economy of energy for transport; Economy measures with existing vehicles; Some slight longer-term developments.# Tables: Breakdown of industry costs (1917); Percentage of direct operating cost attributable to fuel and standing charges.

340 Energy reserves and supplies in the ECE region: present situation and perspectives. Geneva: United Nations, Economic Commission for Europe, 1979. vi, 74p. illus.
Partial Contents: Growth prospects for coal--capital needs and resources, production costs of coal relative to oil and gas; Capital requirement of the energy sector--the development of the capital intensity of the energy sector in the past; Recent events and their effect on energy-related investments; Rising capital intensity of energy production; Macro-economic impact of relatively increased energy-related investments; Improving capital efficiency through international cooperation.

341 Energy: risks, economics, alternatives. (In Energy, vol. 5[1], Winter 1980, pp. 27-28)
Report from the 72nd Annual Meeting of the American Institute of Chemical Engineers, which contends that there are ways of substituting capital and labor for energy, as energy becomes more costly.

342 Energy since the oil war, by J. Grey. (In Astron. & Aeronaut., vol. 13, 1975, pp. 16-27)
Contends, as part of its summary, that economic comparisons of solar energy with fossil fuel and nuclear power plants indicate that ocean thermal plants can cost as much

as $900-1700kWe and still supply cost-comparative electric energy, based on current price for fossil fuels and nuclear power plant capital costs in the $500-1,000kWe range.

343 Energy tariffs and the poor, by the Department of Energy. London: 1976. 31p.
Partial Contents: Electricity and gas domestic tariff structure; Prepayment tariffs; Tariff adjustments and fuel allowances; Concessionary tariffs; Free electricity and gas allowances; Special fuel allowances.

344 Energy: the next twenty years, by H.H. Lansberg. Cambridge, Mass.: Ballinger Publishing, 1979. xxviii, 628p.
Report by a study group sponsored by the Ford Foundation.# Contents: Pt. 1, Energy and the economy; pt. 2, Energy in an international setting; pt. 3, Coal: an abundant resource--with problems; pt. 4, Alternatives to the fossil fuels; pt. 5, Improving the process.

345 Energy used and energy lost: how electricity compares with other energy forms in efficiency and cost, by W.H. Comtous and W.H. Stinson. (In Combustion, vol. 50, Nov. 1978, pp. 26-35)
Paper challenges ideas, and demonstrates that by any valid comparison electrical generation and utilization is a highly efficient process--not wasteful--and more efficient than any practical and acceptable formulation of energy today. Includes cost considerations.

346 Hydrogen energy: economic issues, by R.J. Goettle and R.G. Tessmer. Upton, N.Y.: Brookhaven National Laboratory, 1978. 12p. (CONF-780660-1)
Considers its economy, availability, capital, cost and economic impact.

347 Low energy strategies for UK: economics and incentives, by G. Doyle and D. Pearce. (In Energy Policy, vol. 7[4], Dec. 1979, pp. 346-350)
Presents an economic formulation which they use to determine the conditions under which various home insulation options might become attractive to the private householder.# Headings: Cost effectiveness--a definition; Private vs. public rates of discount; Calculating payback times; Payback time and fuel type; Effect of gas price rises.

348 A policy for more efficient street lighting, by G. Roberts. County of Humberside, [1978?]. [17]p.
Presents various economic measures for reducing annual maintenance costs to improve cost effectiveness of street lighting, which included: 1, Reducing hours of lighting; 2, Switching off lights; 3, Reducing frequency of cleaning and scouting.

349 World energy outlook. Houston, Tx.: Exxon Corporation, 1981. 40p. illus.
Partial Contents: World economic growth; Energy/GNP ratio; World nuclear energy outlook; World coal outlook; World gas outlook; World oil demand and supply; Centrally planned economics energy outlook.

Finance

350 Aims, methods and uses of energy accounting, by F. Roberts. (In Applied Energy, vol. 4[3], July 1978, pp. 199-217)
States the value of energy accounting and defines the subject. Basic methods of energy accounting methods are listed as: input-output table analysis, statistical analysis, and process analysis.# Partial Headings: The effect of energy costs on industrial prices; Energy intensity of capital goods.

351 The airship can meet the energy challenge, by J. G. Vaeth. Astron. & Aeronaut., vol. 12, Feb. 1974, pp. 25-27)
Discusses how it can be designed to move cargo pieces weighing a million pounds and more into difficult-to-reach places at energy expenditures matching resources.

352 Annual report. Washington, D.C.: Federal Energy Administration, 1972+
Gives an overview of the nation's energy supply which includes constraints in finance.

353 Approach to energy financing, by E. L. Kennedy. (In Combustion, vol. 43, Dec. 1971, pp. 35-37)
Paper which draws conclusions at the end of the national Energy Forum, Washington, Sept. 23, 1971. Conclusions dwelt on money, with the contention that financial incentives will stimulate an expanded effort based on existing methods and will both stimulate and widen the use of the great multiplier, technology.

354 Budget estimating hvac jobs, by V. B. Attaviano and N. Wesler. (In ASHRAE J., vol. 13, Sept. 1971, pp. 95-99)
Discusses budget estimates of heating.# Tables: 1, Budget estimating approximate air conditioning load estimates; 2, Budget estimating air conditioning--cost per ton for completely installed system, 1971 price level (New York City Area); 3, Budget estimating air conditioning--cost per ton of selected uninstalled equipment, 1971 price level--US Average; 4, Budget estimates for electrical reheat costs.

355 Capital requirements for the transportation of energy materials based on PIES scenario estimates, by M. J. Kendrick. Arlington, Va.: TERA Inc., 1979. 233p. (Analysis Memorandum AM/E/79-19)

356 China's perplexing energy triangle, by V. Smil. (In Energy Int., vol. 16[6], June 1979, pp. 25+)
China needs to sell oil and coal to finance modernization program. Energy must be exported to raise the capital to finance the growth of energy demand at home.

357 Developing countries: World Bank's energy report. (In Petroleum Economist, vol. XLVII[10], Oct. 1980, pp. 429-430)
Mentions investment in electricity.# Tables: 1, Oil importing developing countries--oil imports, 1970-90; 2, Developing countries--electric power investment, 1981-90; 3, Developing countries--primary energy production and consumption; 4, World Bank programs, 1981-85 (Fiscal year)

358 The dollar problem and disorderly floating exchange rates, by A.T. Ojo. (In OPEC Rev., vol. 3[2], Summer 1979, pp. 86+)

359 Energy and less developed countries in the 1970s. (In Petroleum Economist, vol. 46[10], Oct. 1979, pp. 424-428)
Headings: GNP and population; Balance of payments and external public debt.# Tables: 1, Energy and the oil importing developing countries (OIDCs); 2, Energy and the oil exporting developing countries (OPEC & NOPEC).

360 Energy as a determinant of investment behaviour, by N.D. Uri. (In Energy Econ., vol. 2[3], July 1980, pp. 179-183)
Reviews the theory of investment analysis and presents an econometric examination of the role of energy prices in US investment activity.

361 Energy finance, nuclear finance and the capital markets, by J. Silcock and T. Heyman. (In Nuclear Eng., vol. 20, Mar. 1975, pp. 170-173)
States that energy development represents the world's biggest single financing problem. The need for energy development has coincided with a crisis in the world's capital markets that is as yet unresolved. Nuclear energy represents a major proportion of projected energy investment, and will accordingly face financial problems....# Tables: 1, OECD countries estimated capital investment in OECD energy development, 1974-1985 (assuming oil price $9/barrel); 2, Total OECD energy investment requirement, 1974-1985 comparison with other key figures; 3, Cumulative investment in nuclear capacity; 4, Major electric utilities--projected annual average capital investment...; 5, Share of domestic industrial supply to nuclear power stations ordered in 1965-1968, 1969-1973.

362 Energy in the developing countries: International Bank for Reconstruction and development. Washington, D.C.: World Bank, 1980. 92p.

363 Energy policy in developing countries, by T.W. Berrie and D. Leslie. (In Energy Policy, vol. 6[2], June 1978, pp. 119-128)
Energy investment in developing countries is capital intensive and involves financing problems and difficulties in determining budget priorities.# Partial Headings: Scope of capital requirements; Investment; financing the energy sector; oil, coal, electricity, natural gas.

364 Energy policy, the dollar and the US political system, by R.J. Fiedland. (In Energy Policy, vol. 7[4], Dec. 1979, pp. 295-306)
Explains the fact that the dollar's position as the main reserve currency within the international monetary system prevents the USA from formulating an effective long-term energy policy.

365 Energy strategy and corporate planning. London: Department of Energy, [1978]. 14p. (Energy Commission Paper, 15)
Notes by the Department of Energy, Coal, Gas and Energy Industries. Includes National Coal Board's capital expanditure project: mining (phasing of capital expenditure, 1978/1983). Includes headings under Capital Expenditure; Finance; Financial policies; Pricing policies.

366 Energy thrift investments in France, by A. Weil and D. Probert. (In Applied Energy, vol. 4[2], Apr. 1978, pp. 153-156)
French governmental policy since the oil crisis of 1973 has placed curbing inflation as a higher priority.

367 Foreign investment paces increase, by E. Adams. (In Pet. Eng., vol. 47, July 1, 1975, pp. 23-24)
Considers 1975 as a year of heavy capital investment in the energy industry.# Tables: Estimated capital and exploratory expenditures for facilities and reserves; Sources of capital for exploration and investment outlays, 1965-1975.

368 International financing of energy resources, by A.T. Ojo. (In OPEC Bulletin, June 1980, pp. 18-43)
Paper presented at the Second World Scientific Banking Meeting on International Financing of Economic Development.# Contents: Why international financing of energy resources; Nature and pattern of financing; Special case for international financing of energy in LDCs; Dealing with certain international investment issues.# Tables: 1, Annual petroleum exploration expenditure; 2, Estimated capital expenditures of eight oil companies for 1980 and 1979; 3, Borrowing in international capital markets by purpose of the borrower; 4, Oil exploration drilling density, 1976; 5a, Geophysical survey costs for petroleum exploration, 1977; 5b, Representative exploration costs; 6, Investment in oil and gas by

non-OPEC developing countries estimated requirements, 1976-1985; 7, World Bank lending for energy and power, 1960-1979.

369 Optimal financial choices for the insulation of buildings, by R. J. Wood, P. W. O'Callaghan and S. D. Probert. (In Applied Energy, vol. 5[3], July 1979, pp. 193-204)
A method for determining the economic properties with which various thermal insulation options should be adopted for domestic purposes is described and some results presented.

370 The optimal timing of energy-conserving investment analysis and policy implications, by S. A. Ravid. (In Energy, vol. 5[3], Mar. 1980, pp. 259-270)
Given the cost of equipment replacement and future energy prices, the optimal investment time is calculated.

371 Preparing a case for investment, by B. Locke. Cadogan Consultants, [1979?]. 14p. illus.
Partial Contents: Money flow; Sources of investment funds--venture capital; The case for investment--inflation, capital costs of plant, operating cost of plant, cost improvements, criteria for investment. Concludes that energy conservation is a direct form of cost reduction.

372 Re-examining the nature of ECE energy problem, by A. B. Lorins. (In Energy Policy, vol. 7[3], Sept. 1979, pp. 178-198)
Tables: 1, Approximate marginal capital investment for complete energy system; 2, Approximate prices of delivered energy from selected systems.

373 Report of the working group on energy strategy. London: Department of Energy, 1977. 10p. (Energy Commission Paper, 2)
Partial Contents: Energy forecasts; Financial regimes and pricing policies; Energy policy review.

374 Uncertainties remain in EEC energy policy: Britain's attitudes and oil prices are obstacles, by S. Traill. (In Energy Int., vol. 16[6], June 1979, pp. 28-29)
Includes consideration of funding for the community's R&D programs.

375 Watching the cash flow: energy metering explained, by A. Pipes. (In Energy Manager, vol. 2[2], Mar. 1979, pp. 23-25)
An energy audit is only as good as the new data put into it--the answer is effective metering.

376 What ever happened to the energy crisis? by C. E. Bagge. (In Energy Commun., vol. 3[5], 1977, pp. 459-473)

Contends that capital expansion requirements in the next ten years are estimated to be $18-12 billion in new money.

377 World energy and the EEC, by L. Williams. (In Atom, no. 278, Dec. 1979, pp. 325-327)
Address to the Institute of Nuclear Engineers. Points out that the Council of the European Communities is to spend about £70 million on a second four-year program of energy R&D as announced by the Department of Energy.

Management

378 Accurate temperature control does not mean an expensive controller, by H.A. Wainwright. (In Electrical Rev., vol. 205 [21], 30 Nov. 1979, pp. 25-26)
Contends that the provision of reliable temperature control relies on more than just the accuracy and cost of the controlling equipment.

379 Audit your air system to save energy, by W.M. Kauffmann. (In Power, vol. 119, May 1975, pp. 40, 42)
Presents a detailed calculation for a large industrial plant. Suggests a follow-up of calculation with a check on operation and maintenance. Results will give assurance of lowest-cost air.

380 The brewing industry: energy consumption and conservation in the brewing industry, by P.S. Harris. London: Department of Energy, 1979. 70p. (Energy Audit series, no. 8)

381 Canadian policy in disarray. (In Petroleum Economist, vol. XLVII[12], Dec. 1980, pp. 510-512)
Headings: Costs plus price regime; Boosting federal revenues.

382 Comfort without waste: keeping space heating under control. (In Energy Manager, vol. 3[4], May 1979, pp. 29-31)
Keeping buildings at or around the statutory maximum level of warmth (20° C., 68° F.) costs UK industry about £2,500 million a year.

383 Conserving of energy in buildings, by V.L. Sailor. (In Am. Soc. C.E. Proc., vol. 100[CO 3 no. 10795], Sept. 1974, pp. 295-302)
Analyzes the energy situation in the United States and its relationship to commercial buildings. Factors of importance in the energy budget are: 1, Sources of energy; 2, Anticipated future supplies; and 3, Where and how energy is consumed and how consumption might be reduced through rational action.

384 Cut costs with ROPM, by R.E. Templeton. (In Hydrocarbon Proc., vol. 55, Aug. 1976, pp. 83-85)

The rate of progress method (ROPM) can be helpful during both preparatory and work phases. Contends that a construction organization can properly plan and optimize project cost control even when problems arise to negate it. Includes a table showing project management cost control.

385 A decision support system for energy policy analysis, by R. Pick, A.B. Winston and G. Koehler. Lafayette Institute, Krannert Graduate School of Management, 1980. 26p. (ARO -16231.3-EL)
Proposes a new approach to energy policy analysis.

386 Designing energy-management systems for large industrial power plants, by R.E.J. Putman. (In Power, vol. 119[5], May 1975, pp. 35-38)
Charts: 1, Primary function of an energy-management system for a large industrial power plant is to determine how to run the turbines so demands for process steam and power can be satisfied at minimum cost; 2, Maximum boiler efficiency generally occurs at what is referred to as the normal economic rating, not at full load; 3, Cost of process steam depends primarily on the fuel used; 4, Instantaneous demand limits and incremental power costs are needed to optimize steam and electric production; 5, More efficient utilization of constructed electric power is a continuing goal in industrial plants where success frequency can be measured in thousands of dollars.

387 Energy audits, by B. Buss. London: Electrical Research Association. [1979?]. [20]p.
Explains every auditing as a prerequisite to effective energy management. It is an aspect of financial auditing in that it is concerned directly with costs of energy and its apportionment throughout the entire plant, building, etc.

388 Energy audits--2. London: Department of Energy, 1980. 17p. illus. (Fuel efficiency booklet)

389 Energy conservation in universities: a case study of a Cambridge college, by D. Croghen. Cambridge: Anglian Architects, 1978. [22]p.
Presents the results of a survey in one of the old Cambridge colleges which produced savings of 16 percent balancing fuel price inflation. Over a three-year period the college will reduce energy consumption by 45 percent and cost by 54 percent with full payback of the investment in adopted measures and consultancy.# Partial Topics: Energy audit; Management measures; Cost benefit analysis. Includes completed energy audit forms.

390 Energy in the bank: Natwest's £2m computerised energy package. (In Energy Management, vol. 2[3], Apr. 1979, pp. 37-39)

Looks at National Westminster Bank's new data processing center where energy management is controlled by a £2 million computer. Includes cost considerations.

391 Energy management and waste heat recovery conference. Sponsored by the Institution of Plant Engineers. London: 1980. 1v.
Partial Contents: Making energy management pay; Cutting fuel costs in large to small offices; Energy management in lighting.

392 Energy management in the commercial sector, by G.C. Colwell. London: Marks and Spencer Ltd., [1978?]. 1v.
Explains that energy is a direct charge against the running cost of the business and vital to the manufacturing, distribution and selling process.

393 Energy management: in the engineering industry. London: Institute for Industrial Research and Standards, [1977?]. 1v.
A manual designed to assist the plant manager and accountant to reduce the firm's energy costs.# Partial Contents: Pt. A, Energy measurement, auditing and control--introducing energy management program, auditing control and energy balances, costing of energy, financial appraisal; pt. B, Energy management applications--choice of fuels, units, energy prices, conversion tables.

394 Energy management in the workplace, by M.E. Horsley. (In Plant Engr., vol. 23, Nov. 1979, pp. 20-22)

395 Energy manager--who? Organising an energy management team, by G. Newton. (In Energy Manager, vol. 2[4], May 1979, pp. 19-20)
States that organizing an energy management team means attention to detail and setting priorities at every level of the establishment.

396 Energy managers ... do you still believe in them? by A. Klavenberg. (In Process Engineering, Mar. 1979, pp. 56-57)

397 The energy managers' handbook, by G.A. Payne. Borough Green (Sevenoaks): IPC Science and Technology Press, 1977. x, 147p. illus.
Partial Contents: Management of energy--Senior management, The energy manager, Energy audit, Energy cost, Planning, Regular control.

398 Energy theft: how to find it, how to stop it. (In Pipeline & Gas J., vol. 206[3], Mar. 1979, pp. 34, 36, 38-40, 42)
Special staff report tells how and why US gas utilities are getting "ripped off" to the tune of millions of dollars,

and how they are fighting back to stop the loss of gas and revenue.

399 Energy under the oceans: a technology assessment of outer continental shelf oil gas operations, by D.E. Kash, and others. Norman: University of Oklahoma, Science and Public Policy Program, 1973. xvi, 378p. illus.

Partial Contents: Pt. 3, Policy issues raised by outer continental shelf (OCS) development. Chapter 9, Government management--leasing, planning, cooperation and coordination.

400 Energy use management: proceedings of the International Conference, 1977, Arizona, edited by R.A. Fazzolare and C.B. Smith. New York, Oxford: Pergamon Press, 1977. 2v. illus.

Partial Contents: Vol. 1, Topic B, Industrial sector--Energy accounting and information systems, Fuel and power management; Topic C, Commercial sector; Topic D, Residential sector--Domestic energy management; Topic E, Transportation and communications; Topic F, Agriculture and food; Topic G, Industrialization and development.# Vol. 2, Topic H, Science and engineering; Topic I, Architecture, land use and urban planning; Topic J, The natural environment--energy management in nature; Topic K, The role of government; Topic L, Society and economic aspects.

401 Evaluating investments in energy management. (In Heating/Piping/Air Conditioning, vol. 51[10], Oct. 1979, pp. 69-75)

Concepts and language which need to be understood to sell management.# Headings: Taking stock; Incremental reasoning; payback; Time value of money; Rate of return; Life cycle costing; Tax considerations.

402 Factory services, steam services and insulation, by T.B. Sanson. Pipework & Heating Services (Power Plant) Ltd. [1979]. [9]p.

In any investigation concerning savings that may be made at an industrial plant, it is essential to consider the units as a whole, i.e., Production and power generation, together with usage of same. On this basis, one can define the work or survey under the following headings: 1, Energy savings in the plant; 2, Steam and power requirements, together with the efficiency of same.

403 Federal energy administration project independence blueprint financial task force report--finance: financing project independence, financing requirements of the energy industries, and capital needs and policy choices in the energy industries. Washington, D.C.: Federal Energy Administration, 1974. 1v.

Presents an analysis of the nation's financial capacity to support Project Independence and the impact of the anticipated energy investment upon financial markets.

404 Future developments in energy management, by W.C. Dobie. (In Glass Technology, vol. 20, April 1979, pp. 44-47)

405 Geothermal energy projects: planning and management, edited by L.J. Goodman and R.N. Love. Oxford: Pergamon Press, 1980. xiv, 230p.

406-07 Getting into hot water: district heating looks to the future. (In Energy Manag., vol. 2[4], May 1979, pp. 33+)
Hot water mains could become a familiar sight in towns and cities--the only thing stopping then is cheap gas. Report from the DHA conference.

408 How a consulting firm looks at energy management. (In Heating/Piping/Air Conditioning, vol. 51[10], Oct. 1979, pp. 49-50)
Energy consumption index gives a measure of energy efficiency. A monitoring system called the energy consumption index provides a detailed computer analysis of energy trends.

409 How a contracting firm looks at energy management. (In Heating/Piping/Air Conditioning, vol. 51[10], Oct. 1979, pp. 52-58)
Computerized system analyzes usage, evaluates retrofit options.

410 How an owner's engineer looks at energy management. (In Heating/Piping/Air Conditioning, vol. 51[10], Oct. 1979, pp. 61-66)
Where to look to achieve electricity and fuel savings.

411 How to eliminate heat leaks in your plant, or what to do until the energy doctor comes, by D.N. Wilson. (In Ind. Finishing, vol. 50, Dec. 1974, pp. 12-14, 16)
Describes an approach to eliminating heat waste while awaiting expert help in energy management.

412 International energy management. (In Heating/Piping/Air Conditioning, vol. 51[8], Aug. 1979, pp. 60-61)
An energy reclaim system that reuses cooling system waste heat for preheating service hot water proves effective in energy management.

413 Managing and controlling maintenance, by G. Chavez and H.F. Perla. (In Power, vol. 124[6], June 1980, pp. 108-111)
Argues that a well-planned maintenance-management program can bring organized yet flexible scheduling, improved productivity, and ready access to a data bank.

414 Movement key to prefab-module use, by D.D. Wells. (In Oil & Gas J., vol. 77[42], Oct. 15, 1979, pp. 148, 152, 157, 160, 165-166, 168)

Use of large prefabricated equipment modules can often save construction time and cost. But movement of such units is a key factor in determining the feasibility and advantages of this approach.

415 Net energy analysis: is it a useful tool for energy decision making? by W. Kruvant. (In Energy, vol. 5[3], Summer 1980, pp. 4-5)
Evaluates how helpful net energy analysis is to energy decision making--providing a guide for decision making on energy policy.

416 Outlook for the eighties: summary of 1979 review of energy policies and programmes of IEA countries. Paris: OECD, 1980. 38p.
Partial Tables: Energy trends, 1973-1979; Changes in IEA energy indicators; Global oil balance.

417 Proven ways to save energy in industrial plants, by K.E. Robinson. (In Heating/Piping, vol. 47, May 1975, pp. 58-63)
When it comes to saving energy in industrial plants the only real limitation is imagination.# Chart: Utility cost in a plant before and after energy conservation program was started in 1968 and became truly active in 1969.

418 A quick look at the National Energy Plan, by R.W.A. Legassie and F.I. Ordway. (In Astron & Aeronaut., vol. 15, Nov. 1977, pp. 28-35)
A market-orientated planning study covering the near and not-too-distant future, presages more changes in the fast-moving energy field.# Partial Charts: ERDA (Energy Research and Development Administration) FY75 energy RD&D budget; ERDA FY78 Energy RD&D (Research Development & Demonstration) budget; Increases in ERDA budget authority; FY75-FY78 budget increases for ERDA RD&D: two views; Increases in ERDA budget authority by percentage FY75-FY78.

419 Reduce relief system costs, by T.W. Whelan and S.J. Thomson. (In Hydrocarbon Processing, vol. 54, Aug. 1975, pp. 83-86)
Contends that more emphasis should be placed on the engineering of relief systems because of the potential savings in materials and operating costs.

420 Role of compressors and repowering in energy management. (In Power, vol. 123[3], Mar. 1979, pp. 13)
An editorial comment.

421 Save energy in plant operations, by G.F. Moore. (In Hydrocarbon Processing, vol. 52, July 1973, pp. 67-71)
Efficient practices are the key. Leaky equipment, mis-

use of fuels in plant, and poor design all contribute to the problem. # Partial Charts: Computer data show the dollar loss in steam-condenser ineffiencies; Fuel price escalations increase the need for efficiency.

422 A survey in time saves. (In Energy Manager, vol. 2[9], Nov. 1979, pp. 27, 29)
Industrial surveys reveal high potential for savings.

423 Using consultants, by A. Pipes. (In Energy Management, vol. 2[5], June 1979, pp. 18-19)
A visit by a reputable consultant could save thousands, but what exactly is a consultant? Gives explanation, plus some useful addresses.

424 Wanted: the smaller man; progress report on energy managers groups. (In Energy Manager, vol. 2[2], Mar. 1979, p. 17)
Energy managers group has yet to attract the small to medium firms. The lack of smaller firms in the group in part reflected their interest in energy ... what they don't realize is that energy is almost certainly a much greater proportion of controllable costs.

425 World-wide management development: look at the real ENI, by L.H. Frampton. (In Hydrocarbon Process, vol. 48, Aug. 1969, pp. 84-87)
Gives a candid picture of the structure of "Ente Nazionale Indrocarburi" (ENI) posture and productivity. Prime target of the Italian government is low-cost fuel for the national economy. This has led to the charge that ENI's lack of emphasis on profits is unfair competition.

Pricing

426 Carter unveils crude price decontrol plan. (In Oil & Gas J., vol. 77[15], Apr. 9, 1979, pp. 80-81)
A facet of energy plan which calls for US prices to rise to world level in Sept. 1981.

427 The economics of energy and natural resource pricing: a compilation of reports and headings ..., by The Committee on Banking, Currency and Housing, Ad Hoc Committee on the Domestic and International Monetary Effect of Energy and other Natural Resource Pricing. Washington, D.C.: Gov't Printing Office, 1975. 1v. (variously paged)

428 Energy cartel, coal prices questioned. (In Energy World, vol. 175, Jan. 1, 1971, p. 23)
Does an energy monopoly exist in fact? Are coal prices going to continue upward, or is the market softening?

429 Energy policy, the energy price fallacy and the role of nuclear energy in the UK, by L. G. Brookes. (In Energy Policy, vol. 6[2], June 1978, pp. 94-106)
Rejects the widespread belief that the world energy problem will be solved by rising prices--closing the gap by reducing demand and bringing in new, large, previously over-costly energy sources.# Partial Headings: Energy costs and prices; The economics of energy prices.# Tables show UK energy expenditure as a percentage of GNP, with and without taxation.

430 Energy prices, 1960-1973, by Foster Associates. Cambridge, Mass.: Ballinger Pub. Co., 1974. 1v.
Contains information on energy prices, retail and wholesale prices for primary and secondary energy sources in the United States.

431 Energy pricing, by D. Rooke. London: British Gas Corporation, [1978?]. 3p. (Energy Commission Paper, 10)
Contends that future investment in energy should be profitable; therefore prices for each fuel should ensure that investment to meet future increases in demand can earn a satisfactory return.

432 Energy pricing, by F. Tombs. Electricity Council, [1978?]. 3p. (Energy Commission Paper, no. 9)
A brief statement on energy pricing policy.

433 Energy pricing: an alternative approach, by M. Barnes. London: Electricity Consumers' Council, 1978. 4 leaves. (Energy Commission Papers, no. 11)
States that in a situation where future energy costs are subject to great uncertainty and liable to be significantly affected by the technological change, it is likely that a pricing policy which fixes the price of any particular fuel at a level that will provide a reasonable return on investment and give consumers the right long-term price signals will involve considerable fluctuation in the relative prices of different fuels at different points in time.

434 Energy pricing and inflation in a fixprice-flexprice model, by K. Sharp. (In IEEE Energy, 1978, pp. 35-39)
Develops a model to show how energy price increases are transmitted in other sectors of the economy. Simulations are used to describe how price increases in the monetary sectors eventually nullify the energy price increases.

435 Energy pricing principles. London: Confederation of British Industry, 1979. 6p. (Energy Commission Paper, 16)
Comments by the Confederation of British Industry on Energy Commission Paper, no. 8.

436 Energy: US to phase out crude price control.... (In Petroleum Times, vol. 83, Apr. 15, 1979, pp. 4, 6)

A report on the US President's televised address in which he made a major revision of energy policy by announcing plans for phasing out price controls on indigenous production by September 1981.

437 Energy use trends in industrial countries: implications for conservation, by J. Dunkerley. (In Energy Policy, vol. 8[2], 1980, pp. 105-115)
Topical Headings: Composition of economic output; Differences in energy prices.# Chart: Prices of energy, 1953-76.

438 Historical patterns of residential and commercial energy uses, by E. Hirst and J. Jackson. (In Energy Int. J., vol. 2[2], June 1977, pp. 131-140)
Reviews trends in residential energy use and commercial energy use for 1950-75 period.# Charts: Trends in retail fuel prices, 1950-1975; Household expenditure on fuels, 1950-1975.

439 Impact of energy price increases on low income families. Springfield, Va.: National Technical Information Service, 1976. 1v.
Covers expenditures for gasoline, for home use of piped natural gas, bottled gas, gas or liquid petroleum gas, fuel oil, coal and electricity.

440 Interfuel substitution possibilities: short-term prospects, by N.D. Uri. (In Applied Energy, vol. 4[4], Oct. 1978, pp. 251-260)
Applies a translog price possibility frontier in order to measure the extent of interfuel substitution effects in the electric power industry in the United States. The results suggest that relative changes in fuel prices have significant effects on fossil fuel consumption.

441 Modelling UK energy demand to 2000, by S.D. Thomas. (In Energy Policy, vol. 8[1], Mar. 1980, pp. 17-37)
Appendices: 1, Calculation of initial data. Table 8, 1976 market share, fuel price and efficiencies: domestic sector; 9, 1976 market shares, fuel prices and efficiencies; production sector; 10, 1976 market share and specific fuel consumption transport sector; 2, Fuel pricing data. Table 11, Pricing data for coal, gas, oil and electricity; 12, Electricity production costs; capital charges (1976 oil price = 1.0); 13, Electricity products costs: overheads (1976 oil price = 1.0); 3, Detailed working assumption for the energy balances. Table 14, Assumptions used for converting from therms supplies to primary fuel....

442 The pattern of domestic energy consumption and the growth of prices in relation to consumers' income and expenditure, 1966-1967, by F. Tombs. London: Electricity Council, [1978]. 6p. (Energy Commission Paper, no. 21)

Reviews the facts about the burden of fuel costs and how much people have to pay out of their incomes for their fuel. Includes data on household expenditure and the prices of domestic fuels, together with a comparison of figures for previous years.

443 Policy on energy pricing, by M.G. Webb. (In Energy Policy, vol. 6[1], Mar. 1978, pp. 53-65)
Contends that pricing of particular fuels, such as electricity, gas and oil, should be considered in terms of the development pricing policies for the ernegy sector as a whole.# Partial Headings: Marginal cost pricing; Measures of marginal costs; The differentiated tariffs; Financial targets; Income distribution. Table shows energy price elasticities for the USA, 1968-1972.

444 Price sensitivity of petrol consumption and some policy implications: the case of the EEC, by G.J. Kouris. (In Energy Policy, vol. 6[3], Sept. 1978, pp. 209-216)
Using econometric and statistical techniques, this study focuses on conserving energy by taxing petrol. If the purpose of taxation on petrol prices is to raise public revenue, then the insensitivity of petrol demand to rise would guarantee a good tax yield.# Charts: Petrol consumption as a function of price and real disposable income; EEC petrol consumption growth rates.# Partial Tables: Short run income and price elasticities (income and petrol consumption per capita); Petrol consumption income and price indices EEC (1963=100).

445 Real price of energy, by F.B. Dent. (In Am. Gas Assn. Mo., vol. 56, Apr. 1974, p. 25)
A short article about American energy policy. "Of all the wrong approaches to energy policy, none has been more illusory or counter-productive than the federal government's 20-year attempt to set a just and reasonable price for natural gas at the well."

446 Security price index record. New York: Standard and Poor Corp., 1962+
Annual. Presents time series data on the daily and weekly stock price indicators for petroleum and coal as well as utility stocks.

447 TUC statement. Trades Union Congress, 1977. 9p. (Energy Commission Paper, 4)
Under pricing and finance, the TUC contends that a part of the North Sea oil revenues should be used to finance investment in the coal and nuclear industries, as well as wave and solar power.

448 What are prices of gas and oil from coal? by W.L. Nelson.

(In Oil & Gas J. Tech., vol. 77[17], Apr. 23, 1979, pp. 67)

Question asked was: What the costs of gas and oil by coal conversion will be? Can answers be used in estimating such costs.

449 What is fair price for steam from a neighbor? (In Power, vol. 117, Apr. 1973, pp. 134-135)

Gives solutions as follows: Buy own oil-fired boiler; Fixed unit price is unfair; Contractor's bid gives pricing start; Put details into contract; Talk to utilities commission; Include pipeline costs, too.

450 Where aerospace can serve afresh: paths to energy independence, by W.M. Hawkins. (In Astron. & Aeronaut., vol. 16, Feb. 1978, pp. 32-36)

Contends that if government for its own purposes can pick up the difference between Near East oil prices and the artificially controlled prices in the US, it can certainly afford to experiment with systems of marginal cost that promise eventual economies or ultimate freedom from dependence on tradtional energy sources.# Charts: 1, US energy consumption and potential sources; 2, Sources of petroleum to fill demand; Relative costs of aircraft fuels. Includes table showing comparison of liquid-hydrogen-fueled and conventional cargo transport.

Statistics

451 Air transport facts and figures. Washington, D.C.: Air Transport Association of America. 1940+

Annual. Presents a statistical review which contains energy related data.

452 Domestic oil reserves forecasting method, regional potential assessment, by J.A. Momper. (In Oil & Gas J., vol. 77[33], Aug. 13, 1979, pp. 144-146)

Tables: 1960 forecast; comparison: 1960, 1964, 1968, 1972 forecasts; 1978 forecast; comparison; 1960 vs. 1978 forecasts; Petroliferous provinces of the US; Producing regions of the USA; Rocky Mountain producing region.

453 Energy balances of OECD countries ... 1974/1978. Paris: Organisation for European Cooperation and Development 1980. xiv, 166p.

Based on data recently published in "Energy Statistics, 1974-1978." Gives energy balance sheets for OECD member countries total energy requirements, growth triangles, and energy consumption/economic.

454 Energy banks for small villages, by I.H. Usmani. (In Bull. Atomic Scientist, vol. 35[6], June 1979, pp. 40-44)

Contends that on the whole the energy situation in the developed countries is currently manageable even though there may be painful long-term adjustments along with acute shortages of limited duration. # Partial Tables: 3, Financing of rural electrification demonstration program of 300 small (500 persons) and remote villages, 1982 to 1985.

455 Energy, GNP and causality: a statistical look at issue, by N. D. Uri. (In Energy Communications, vol. 6[1], 1980, pp. 1-15)
Discusses the relation between energy consumption and economic activity as bidirectional. # Tables: Summary of the regression; F-Tests on four future coefficients; Lag distribution from the regression. # Chart: Cumulative periodogram of residuals.

456 Oil and gas are not enough, by E. Adams. (In Pipeline & Gas J., vol. 197, May 1970, pp. EM5-8)
Discusses the state of the oil and natural gas industry. # Chart: Earnings of petroleum companies. # Tables: 1, Financial statistics of US petroleum companies; 2, Financial statistics of US natural gas companies.

457 Refiners face challenges of 70s, by J. D. Wall. (In Hydrocarbon Process, vol. 49, Dec. 1970, pp. 67-73)
Partial Topics: Energy demand; Location of refining centers; Production distribution; Competition. # Charts: 1, Energy consumption in the Free World; 2, Energy sources in the Free World; 3, Oil consumption in the Free World; 4, Refining growth in major world areas, 1960-1980; 5, Refining growth in the Eastern Hemisphere area, 1960-1980; 6, Refining growth in the Western Hemisphere area, 1960-1980; 7, Western European and US refining yields, 1938-1969.

458 Short-term forecasting of crude petroleum and natural gas production, by N. D. Uri and S. P. Flanagan. (In Applied Energy, vol. 5[4], Oct. 1979, pp. 297-310)
Details the Box-Jenkins approach to forecasting time series and applies it to short-term natural gas production and crude petroleum production in the United States.

Taxation

459 Carter spells out "windfall" tax proposal. (In Oil & Gas J., vol. 77[18], Apr. 30, 1979; pp. 124-125)
A request to Congress to impose complicated level that would keep most existing US crude under price control until 1990. # Table: Refiners crude costs during decontrol, 1979-1982.

460 The effective tax rates of extractive industries and the case

for a geothermal resource tax incentive, by J. McNamara. (In Geothermal Energy, vol. 6[6], June 1978, pp. 11-18)
Tables: A, Comparative income tax returns--1974 (IRS, 1976); B, Corporation income tax returns--1973 (IRS, 1975); C, Corporation income tax returns, 1972 (IRS, 1974); D, Tax treatment of income from oil and gas; E, Changes in reported oil company profits (1967=100)

461 Energy taxes: a method for financing the national health insurance, by J. Kraft. (In Energy, vol. 4[3], June 1979, 429-438)
Shows that energy tax would generate sufficient revenues to finance a national health care program.

462 Estimates of energy non-resource costs: energy taxes and subsidies--analysis report. Washington, D.C.: Department of Energy, 1979. 29p.
Establishes quantitative estimates of selected taxes and subsidies that impact the cost of producing and consuming different energy sources.

463 Senate poised to soften excise tax blow. (In Oil & Gas J., vol. 77[29], July 16, 1979, pp. 34-35)
Attempts by lawmakers to weaken House bill with amendments.

464 Windfall profits tax: finally a reality. (In Energy, vol. 5[2], Spring, 1980, pp. 4-5, 24)
Presents highlights of the windfall profits tax. Table shows estimated windfall tax assessment.

IV. FUELS

Costs

465 Air preheater can help reduce fuel costs; data sheet, by R.G. Schwieger. (In Power, vol. 119, Jan. 1975, p. 12)
Presents a chart which shows the potential gross saving in annual fuel cost that can be achieved by installing a regenerative-type air preheater on a conventional industrial boiler.

466 The allocation of transport fuels to minimise costs, by B.D. Armstrong. (In Transport & Road Research Laboratory Digest of Report LR956, 1980. 2p.)
Concerned with a time in the future when natural oil--and natural gas--become substantially more expensive than oil and gas made from coal. Assesses the total resource cost to the nation of a given allocation of a fuel and highlights the way in which the total cost changes in electric or liquid fueled vehicles are used.

467 Alternate fuel cost detail. (In Oil & Gas J., vol. 77[38], Sept. 17, 1979, pp. 92, 94, 97)
Contends that much of the technology for reducing a variety of alternative fuels is available, but the economics of developing most of these alternative sources is tenuous at best.# Tables: Effect of coal costs on price of liquids; Synthetic gas costs ...; Basic economic conditions from coal costs ...; Coal liquefication costs ...; Syncrude project expenditures ($ billion); Coal conversion economics.

468 Alternative fuels, by the Boeing Company. (In ITA Bull., 36E, Nov. 1980, pp. 849-856)
Includes a chapter on alternative fuel costs and a table showing energy capital requirements in billions of US dollars.

469 Are we overlooking the fuel cells? by S. Baron. (In Elec. World, vol. 174, Dec. 1, 1970, pp. 44-46)
Contends that fuel cells offer a new direction in power generation that could be attractive economically while at the same time making a major contribution to reducing environmental effects.# Table: Costs to gasify coal (coal cost not included).

470 Aviation fuel: supply prospects to the year 2000. Montreal: ICAO. Air Transport Bureau. (In British Airways Air Safety Review, 1980, pp. 2-8)
An analysis of the fuel situation over the next 10-20 years. ICAO provides a glimpse of what can be expected by the air transport industry regarding costs and the availability of aviation fuel....# Partial Tables: 2, Trend in crude oil prices, based on a market price index, developed by the organization for Economic Corporation and Development, 1961-1974; 3, Average fuel expenses and total operating expenses per available ton-kilometer for world's scheduled airlines, 1961-77; 5, Average aviation fuel prices for international services of IATA carriers, 1973-1978.

471 Breakeven fuel cost spread widens, by R. R. Bennett. (In Elec. World, vol. 183, Mar. 15, 1975, pp. 120-121)
Climbing capital costs coupled with high interest rates militate against nuclear and coal-fired generation.# Tables: Capital needed for nuclear and coal-fired units exceeds requirements for other forms of generation. This results in a breakeven fuel-cost spread that responds to interest rate changes; Investment estimates for selected sizes; How acceptable cost differentials have increased.

472 Compare power plant fuel costs: data sheet, by C.C. Richard. (In Power, vol. 117, Dec. 1973, p. 14)
Presents a chart which shows the difference in fuel costs between alternative generating plants.

473 Cost and quality of fuels for electric utility plants. Monthly report data for November 1978. Washington, D.C.: Energy Information Administration, 1979. 63p.
Presents cost analysis and prices.

474 Cost controls called double-edged sword, by H. Gregory. (In Aviation Week, vol. 100, Apr. 15, 1974, pp. 36-37)
Contends that cost control as a way of profitability for airlines is a popular theme in the financial community, but in the viewpoint of at least one airline the realities of life as a service industry have not changed fundamentally with the fuel crisis.

475 Crude costs seen near parity with alternate fuels. (In Oil & Gas J., vol. 77[45], Nov. 5, 1979, pp. 54-55)
Comment on what H. C. Kauffmann said at a business session, that price increases by oil exporting countries have pushed crude costs close to parity with alternative fuels....

476 Cut those fuel costs, by P. Abbott. (In Civ. Eng., Sept. 1979, pp. 16-17)

477 Cutting light weight aggregate fuel costs, by N. W. Biege and

S.M. Cohen. (In Rock Prod., vol. 77, Nov. 1974, pp. 50-51+)

478 Economic assessment of the utilization of fuel cells in electric utility systems, by W. Wood, and others. (In Conf. New Opinions in Energy Technology, by AIAA/EEI/IEEE, 1977. pp. 51-61)
Benefits of fuel cells' unique characteristics are separately quantified and the relationships between market penetration, fuel price and capital cost are analyzed utilizing reliability, production cost, and optimum generation mix methods.

479 The energy problem: its effect on aircraft design: pt. 2, The effect of fuel cost, by W. Tye. (In Airc. Eng., Apr. 1980, pp. 2-4)
Topical Headings: What decides cost and price; Effect of energy costs on the pattern of living; The effect of energy cost on air transport; Fuel crisis and aircraft operating costs.

480 Ethanolic fuels from renewable resources in the solar age, by H.P. Gregory and T.W. Jeffries. New York: Columbia University, Department of Chemical Engineering and Applied Chemistry, 1979. 34p.
Describes how ethanol, other than liquid fuels, can be produced from cellulosic biomass at prices that can compete with petroleum-derived materials. Contemporary membrane technologies can already effect major savings in process costs.

481 FCC power recovery save $18 million at refinery. (In Oil & Gas J., vol. 77[47], Nov. 19, 1979, pp. 164-166, 168)
From 1973 through 1978 a power recovery train has saved Sun Petroleum's Toledo refinery an estimated $18 million in fuel costs. Includes table showing comparison of power costs.

482 Fixed charges, fuel share blame for record energy cost. (In Elec. World, vol. 184, Nov. 15, 1975, pp. 44-47)
Sharply higher fixed charges coupled with soaring fuel prices boosted total busbar-energy costs for 1974 to the highest level ever reported.# Tables: Summary of key factors reveals pattern of power cost rise over 10-year period; Complete data for 18 modern steam stations show range of total energy costs during 1974; Taxes cause fantastic markup for energy costs; Price hikes make oil costliest energy source; Soaring fuel prices cause massive boost in total busbar-energy costs.

483 Fuel conservation saving cents makes sense, by T. Wilson. (In Touchdown, 1980, no. 1, pp. 24-27)
Indicates various areas where careful routine maintenance

can help avoid fuel wastage and also identify the qualities of fuel likely to be involved to aid assessment of the cost-effectiveness of maintenance action.

484 Fuel consumption and cost. Washington: Gov't Print. Office, 1976+
Monthly. Gives data on fuel consumption and costs of individual trunk and supplemental air carrier. Details include fuel consumption and costs, average price per gallon, and percentage change in price per gallon.

485 Fuel cost and consumption: twelve months ended December 31, 1979 and 1978. Washington, D.C.: Civil Aeronautics Board, 1979. [12]p.
Tables: Highlights; Trunks and local service carriers; Trunk carriers; Local service carriers; All cargo carriers; Charter carriers; Alaskan and Hawaiian carriers; Other carriers.# Gives statistics for all certificated carriers--domestic and international operations.

486 Fuel costs hinder Olympic's return to profitability, by H. Lefer. (In Air Transport World, Nov. 1979, pp. 78-80, 83-84, 86)
Describes the optimism of management that new streamlined organization, improved labor relations and a billion dollar re-equipment plan will pay off.

487 Fuel costs seen shaping future designs, by B.M. Elson. (In Aviation Week, vol. 102, Apr. 28, 1975, p. 105)
Contends that fuel prices will become a major factor in the future design of engines, wings, avionics and the air traffic control system.

488 Fuel costs shift charter focus, by L. Doty. (In Aviation Week, vol. 100, Jan. 21, 1974, pp. 26-27)
Scheduled carriers curtail sales efforts but supplementals foresee few gains in any consequent easing of competition.# Chart comparisons for 1968 and 1972 show that US share of international charter seats has increased substantially.

489 Fuel processes must consider energy costs. (In Chem. & Eng. N., vol. 53, June 2, 1975, pp. 18-19)
Contends that energy crunch may force fuel production system designers to base crucial decisions on energy rather than monetary balance.

490 Fuel savings for first diesel electric supply vessel will run $1 million a year. (In Ocean Industry, vol. 14[2], Feb. 1979, pp. 115+)

491 Fuelwatch: IATA cites worldwide impact of soaring fuel costs, by A. Eddy. (In Aircargo Magazine, Aug. 1979, pp. 36-37)

A review of how fuel shortages and rising costs are affecting various cargo operators.

492 How European producers offset rising fuel costs, by O.F. Feddersen. (In Rock Prod., vol. 77, May 1974, pp. 67-69+)

493 Industrial boilers: fuel switching methods, costs, and environmental impact, by J.M. Burk and M.D. Matson. Austin, Tx.: Radian Corporation, 1978. 182p.
The potential for existing boilers to switch fuels from natural gas or oil to oil, coal, or a coal-derived fuel is examined. The technical aspects of switching fuels, capital and the operating costs, and the environmental effects are also discussed.

494 Jet fuel: counting the cost, by B. Rek. (In Flight Int., May 12, 1979, pp. 1566-1567)
Airlines belonging to the International Air Transport Association (IATA) implemented fuel-price surcharges on international fares and cargo rates ranging from five to seven per cent, depending on the geographical area. This move represents a reaction upheaval in world oil markets, and caused jet oil prices to rise dramatically as well as shortages in various parts of the world.# Table: Fuel cost increase: effect on operating costs.

495 Lack of money, not fuel, causing energy crisis, by E. Drake. (In Engineer, vol. 237, Sept. 27, 1973, p. 11)
Reckons that there will be a requirement for £35 million a day, or £200,000 million over ten years to develop existing proven reserves to meet the rising demand for oil.

496 Low-cost fuels coming for gas turbines, by S.M. Delorso and G. Vermes. (In Elec. World, vol. 176, Sept. 15, 1971, pp. 70-73)

497 Low fares don't cost le$$, by B. Rek. (In Flight Int., Dec. 22, 1979, pp. 2083-2084)
Contends that fuel price has become the most worrying item in airline cost breakdowns.

498 The military utility of very large airplanes and alternative fuels, by N.T. Mikolowsky, L.W. Noggles and W.L. Stanley. (In Astron. & Aeronaut., vol. 15, Sept. 1977, pp. 46-56)
Contends that a million-pound-plus airplane using conventional jet fuel should be more cost effective and energy effective than any of today's large airplanes in a variety of military applications.# Partial Tables: VLA life-cycle cost estimates; Cost estimates for the synthetic chemical fuels; Relative cost of energy effectiveness for strategic aircraft missions.

499 Monograph speeds fuel and energy cost conversions, by L.H.

Dierdorff. (In Pipeline & Gas J., vol. 201, Mar. 1974, pp. 27-28)
Discusses special monograph which makes it possible to calculate equivalent costs of many energy sources to determine what really is expensive and what is economical.

500 Non-fossil fuels raise costs, by K. R. Williams and N. V. L. Campagne. (In Hydrocarbon Process, vol. 52, July 1973, pp. 62-64)
Estimates for future systems show costs that are multiples of today's fuel costs.

501 Post-control fuel costs weave old upward pattern, by K. W. Bennett. (In Iron Age, vol. 213, May 20, 1974, pp. 25-26)
Contends that with prices at horrendous levels, buyers feel the oil situation is a classic model of the havoc that big government can wreak.

502 Preheaters equal lower fuel costs, higher production, by J. L. Robertson. (In Rock Prod., vol. 79, June 1976, pp. 55-58+)

503 Program assesses fuel cost and security, by P. J. Peschon. (In Elec. World, vol. 177, May 1972, pp. 42-43)
A digital computer program for simultaneously computing the production costs and assessing the security status of generating units has been used successfully by utilities in Europe since 1966. This program, based on the probabilistic approach, has been refined by Systems Control Inc., under the name of Procos. The modified program is designed to offer several advantages.# Tables: 1, Characteristics of units in ... Systems; 2, Output and fuel costs during week 40.

504 Questions and answers on cutting fuel costs, by W. Short and others. London: Graham & Trotman Publishers, 1975. [8], 104p.
Contents: Questions about fuel supply; Questions about fuel usage; Questions about boiler efficiency; Questions about boiler operation: Questions about insulation; Questions about conservation; Questions about government aid.

505 Rating aircraft on energy, by D. V. Maddalon. (In Astron. & Aeronaut., vol. 12, Dec. 1974, pp. 26-43)
Designing aircraft with fuel consumption in mind is nothing new. Any commercial aircraft development program has as one of its design goals, moving the maximum amount of disposable payload (payload plus fuel load) at the least possible energy cost. The present sharp increase in price of fuel has made this consideration even more important.# Partial Tables: Intercity modal energy intensity comparison, 1970; US airline fuel consumption, 1950/1952; US aviation energy consumption.

506 Recursive estimation of incremental cost curves, by F.J. Taylor. (In Comput. Elect. Eng., vol. 4[4], Dec. 1977, pp. 297-307)
Contends that the economic scheduling of power generation among fossil fuel plants has historically been based on incremental cost curves which relate the rate of change of cost to change in power delivered to transmission system.

507 Remove sulfur from fuel oil at lowest cost, by J.H. Blume, D.R. Miller and L.A. Nicolai. (In Hydrocarbon Processing, vol. 49, Sept. 1969, pp. 131-137)
Costs are compared for several routes to low sulfur fuel oil.

508 Symposium on clean fuel from coal, 1973. Chicago, Ill.: Institute of Gas Technology, 684p. illus.
Partial Contents: Economic evaluation and process design of a coal oil gas refinery; The cost and commercialization of gas and liquids from coal.

509 Synthesized gaseous hydrocarbon fuels. Washington, D.C.: Department of Energy, Federal Energy Regulatory Commission, 1978. xiv, 291p. (DOE/FERC-0008)
Report to the Federal Energy Regulatory Commission by the Technical Advisory Task Force on ...# Partial Contents: Economics of synthesized gaseous hydrocarbons--capital requirements and annual operating costs; Economic evaluating methods.

510 The threatening crisis, by B.A. Rahmer. (In Petroleum Economist, vol. XLVII[2], Feb. 1980, pp. 49-51)
An introductory chapter to "Fuels for the Future," by B.A. Rahner, a new study of world energy alternatives.# Partial Heading: Production costs of alternative energy in terms of 1979 dollars per barrel of oil equivalent.

511 Uranium vs. fossil fuels: an energy source, by E.S. Bell and F. Angebrandt. (In Can. Min. & Met. Bull., vol. 62, Sept. 1969, pp. 972-979)
Considers alternative fuels for electric power generation in Canada.# Tables: 1, Comparative generation costs (4 unit plants: at 7 percent interest rate); 2, Comparative generation costs (4 unit plants; at $5\frac{1}{2}$ percent interest rate); 3, Comparative fuel costs; Electricity demand and supply.# Partial Charts: Variation of total capital cost with unit size; Total annual cost (in $ per kw) for nuclear conventional thermal gas turbine stations--7 percent cost of money; Total annual cost (in $ per kw) for nuclear conventional thermal and gas turbine stations--$5\frac{1}{2}$ percent cost of money.

512 US crude costs leap. (In Petroleum Economist, vol. 46[12], Dec. 1979, pp. 514)
The cost of foreign crude oil landed in USA took its

predictable leap in the third quarter, to $22.13/barrel from $17.24 in the second quarter and $14.25 in the like 1978 quarter.

513 Water turbine technology for small power stations, by T. Salovaara. (In Energy, vol. 5[1], Winter 1980, pp. 23-24)
Interest in the potential of small hydropower stations is growing due to escalating fuel costs, and technology is changing due to rising construction costs and different needs. Describes a Finnish water-turbine experience. Presents a graph of typical plant cost breakdown.

Finance

514 Accounting for intangibles in a present worth comparison of advanced power generation alternatives, by B.D. Pomeroy and J.J. Fleck. (In IEE Power Eng. Soc. Paper from the Joint Power Gen. Conf., 1978. 8p.)
Accounts for inflated capital, fuel, and labor costs, by using a modified present worth analysis of revenue requirements.

515 Airlines detail fuel problems, by R.P. Hudock. (In Astron. & Aeronaut., vol. 12, Apr. 1974, pp. 8-11)
Discusses representations made by the nation's airlines about their views on the fuel shortage. They chorused higher costs, lower profits, and an inadequate federal program. One answer put forward was to have Congress move to control the price that the airlines must pay for fuel, in order to protect the industry from financial catastrophe and the traveling public from exorbitant fare increases.

516 Family expenditure survey: expenditure on fuels, 1977. London: Department of Energy, 1977. 37p.
Partial Contents: Average expenditure on fuel--its variation with household income; Fuel expenditure and income; Fuel expenditure and type of dwelling; Fuel expenditure and household composition.# Partial Tables: Fuel by income; Fuel by dwelling; Fuel by household composition; Fuel by pensioner households; Proportion of central heating installations by fueling and household income, 1977; Motor fuel by income; Payment method by income.

517 Family expenditure survey: expenditure on fuels, 1978. London: Department of Energy, Economics and Statistics Division, 1980. 43p. illus.
Partial Contents: Average household expenditure and expenditure on fuel; Average expenditure on fuel ...; Fuel expenditure and income for consuming households; Fuel expenditure and type of dwelling; Fuel expenditure and household composition; Method of payment for electricity and gas.

518 Impact of government subsidies on market penetration of synthetic fuels, by A. Ezzati. (In Energy Policy, vol. 6 [3], Sept. 1978, pp. 196-208)
Assesses the impact of the various levels of government subsidies on the degree of US dependency on imported energy for the period 1977-2001 and on the reduction of energy bills under the "base case," conservation and most likely scenarios using the Gulf/SRI US energy modes.# Charts: Cumulative production of energy sources as a foundation of marginal cost; Synfuels production under various government price subsidy programmes--"most likely" scenario.# Partial Tables: Forecasts of energy supplies and prices (base cases, no price subsidy); Annual and cumulative costs to government for conservation and "most likely" scenarios of 50 cents subsidy programs; Total energy bill with price subsidy programs.

519 Research on synthetic fuels intensified, by W.C. Wetmore. (In Aviation Week, Dec. 10, 1979, pp. 37-39, 42-43)
Contents: Financial investment; Required investment; System costs; Gate costs.

520 A review of research on variations in fuel expenditure, by J. Bradshaw, A. Briggs and L. Lewis. York: University of York, Department of Social Policy Research Unit, 1978. 34p.
Review carried out on behalf of the Electricity Consumer Council.

521 Single family households: fuel oil inventories and expenditures --national interim energy consumption survey. Washington, D.C.: Department of Energy, 1979. 24p.
Conducted surveys which include information on the households' expenditures for fuel oil.

522 US mergers in the melting pot, by B. Rek. (In Flight Int., vol. 115, Feb. 24, 1979, pp. 564-566)
Looks at the reasons for this movement, which arises out of delegation and fears of deteriorating finances. Cites the intended merger some years ago between Pan Am and TWA when the former was badly hit by recession and the fuel price rise, and recorded severe losses.

Management

523 Aviation fuel: which way do we go? (In Astron & Aeronaut., vol. 13, Oct. 1975, pp. 12)
Contends that there are basically two groups of alternate fuels: the cryogenic fuels require a major upheaval in aircraft design and in airport facilities. This poses formidable problems beyond capital resources problems. Presents a table of alternate fuels--properties and cost.

524 Boeing designs fuel economy into 727, by R.G. O'Lone. (In Aviation Week, vol. 101, Aug. 5, 1974, pp. 24-26)
Improvements made on the latest stretched 727 offering, the 727-300, in order to provide the improved economic and fuel consumption characteristics demanded by today's airline operating environment.

525 Burning tomorrows fuels.... (In Power, vol. 123[2], Feb. 1979, pp. S1-S24)
A special report which makes comparisons of various fuels--gases, liquids, solids and slurries.

526 Cessna cites light aircraft fuel saving, by E.J. Bulban. (In Aviation Week, vol. 101, Aug. 19, 1974, pp. 69-70)
Cessna's 1975 light single-engine line emphasizes fuel economy and speed advantages of these airplanes over surface transportation to capitalize on trends the company has seen in growing use of business aircraft since the energy crisis.

527 Challenge to aviation: improve fuel economy, by F.E. Moss. (In Automotive Eng., vol. 83, Aug. 1975, pp. 38-39)

528 Daily fossil fuel management, by J.W. Lamont and others. (In IEEE Power Ind. Comput. Appl. Conf., 1979, pp. 228-235)
Contends that the problem of daily fuel management for electrical generators has become more complex with the use of fuels of different types, from different sources and costs at each generating unit.

529 Economic fuel, generation and environmental implications of load management and conservation, by F. Okose and others. (In IEEE Power Eng. Soc. Prepr. Summer Meeting, 1979. 8p.)
Demonstrates the considerations and methods required for the evaluation of load management and conservation programs and reports the results in terms of cost, generation requirement, fuel consumption and environmental pollution.

530 The efficient use of energy combustion control, by C.W.E. Hardy. Dorset: Hamworthy Engineering Ltd., [1979?]. [29]p.
Outlines the use of combustion control equipment, designed to control the air-to-fuel ratio on boiler plant, and by so doing provide a useful guide to the energy manager and others interested in fuel. Savings which are to be found in exercising control in this direction.

531 The energy problem: its effect on aircraft design, by W. Tye. (In Aircraft Eng., Mar. 1980, pp. 9-11)
The first of a four-part series of articles. Part one is on supply and demand.

532 Forward planning of fuelling outages, by T. L. Hayslett, J. R. Ratcliff and T. L. Robertson. (In Nuclear Eng. Int., vol. 24, Apr. 1979, pp. 47-52)
As a result of an extensive survey, the Tennessee Valley Authority reckons it can save nearly $1 million per reactor by moving from an annual to an 18-month fuel cycle.# Table: Once-through fuel cycle--typical component costs.

533 Fuel economy and the difficulties in providing and holding dollar savings. (In Diesel Equip. Supt., vol. 53, June 1975, pp. 36-40)
A panel discussion.

534 Fuel efficiency pivotal in fleet expansion plans, by J. Ott. (In Aviation Week, Mar. 3, 1980, pp. 181, 183)
Tables: US airline operating expenses per average ton-mile; Air freight traffic and revenue.

535 Fuel savers in the cockpit, by R. Whitaker. (In Flight, Feb. 28, 1981, pp. 563-564, 567-568)
As fuel costs have soared and availability has become less certain, aircraft operators grasp at any method of reducing the gallon count.

536 Fuel savings = big dollars at McDonnell Douglas, by M. Ogden. (In Diesel Equip. Supt., vol. 54, Oct. 1976, pp. 80-87)
Discusses the concept of total equipment management which slashes accident costs and maintenance expenses as well as achieving amazing fuel economies.

537 Fuel starvation, by J. M. Ramsden. (In Flight Int., Mar. 29, 1980, pp. 988-990)
Gives a first-principles background to fuel saving by improved technology and flight management.

538 Japanese refinery fuel savings techniques explained. (In Oil & Gas J., vol. 77[38], Sept. 17, 1979, pp. 102, 105-106)
After the increase in crude oil prices which began in 1973, Japanese refiners began energy conservation programs to reduce costs.

539 Linear programming techniques can cut fuel management costs, by M. L. Brown and others. (In Elec. World, vol. 17, Mar. 15, 1972, pp. 52-53)

540 Options for reducing fuel usage in textile finishing tenter dryers, by W. H. Hebrank. (In Am. Dyestuff Rep., vol. 64, Apr. & May, 1975, pp. 32, 34; 44-46)
Expected savings are demonstrated through sample calculations for a hypothetical dryer oven.

541 Reactor and fuel: an integral system, by R. B. Richards and

S. Leng. (In Elec. World, vol. 173, Apr. 20, 1970, pp. 27-28)
An understanding of this interrelationship is important to economical fuel management and procurement practices.

542 Spent fuel management: report/INFCE Working Group 6. Vienna: International Atomic Energy Agency, 1980. 113p.

Pricing

543 APPA study trains heavy guns on fuels industry. (In Elec. World, vol. 175, Apr. 15, 1971, pp. 28-29)
A study of fuel supplies and prices, disclosed earlier behind closed doors to American Public Power Association Executives, was publicly aired at a Tennessee Valley Public Power Association meeting.

544 Aviation fuel: a major growth sector. (In Int. Petr. Times, Apr. 1, 1980, pp. 18-20)
Contends that despite rising oil prices and air fares, the public and business community appear to have a voracious appetite for air travel, and price demand plateaux are impossible to predict. Table shows aviation fuel prices, 1934-1979.

545 Aviation fuel and the airlines, by P. Maurice. (In Interavia, 11/1980, pp. 1008/9)
The price of aviation fuel has multiplied by a factor of between seven and ten times.# Charts show average fuel in US cents per US gallon for an airline; and fuel cost as percentage of direct operating costs for financial years 1972-1980.

546 Aviation fuel: supply prospects to the year 2000. (In ICAO Bull., Dec. 1979, pp. 11-16)
An analysis of the fuel situation over the next 10-20 years, which provides a glimpse of what can be expected by the air transport industry regarding costs and the availability of aviation fuel.# Tables: 1, Comparison of world energy as a whole and oil consumption, 1965-1978; 2, Trend in crude oil prices, based on a market price index, developed by the Organization for Economic Cooperation and Development, 1961-1974; 3, Average fuel expenses and total operating expenses per available ton-kilometer for world's scheduled airline; 4, World resources of principal fossil fuels...; 5, Average aviation fuel prices for international services of IATA carriers, 1973-1978.

547 A fresh look at aviation fuel prices: recent and projected prices underscore the requirements placed on designers to incorporate the latest technology into new airplane programs

to reduce fuel consumption. (In Astron. & Aeronaut., vol. 18[3], Mar. 1980, pp. 34-35)

Charts: 1, New estimate of economic kerosene jet-fuel prices; 2, High-low projections of the impact of fuel prices on 1987 Direct Operating Cost (DOC).

548 Fuel outlook dictating technical transport research, by W.C. Wetmore. (In Aviation Week, vol. 101, Oct. 28, 1974, pp. 52-53)

With jet fuel available in reasonable quantities but at prices at least double what they were in mid-1973 and almost certain to go higher, and with the anticipated low level of domestic petroleum reserves on hand at the turn of the next century, NASA is pushing energy-efficient aircraft concepts, as a means of reducing operating costs.

549 Fuel price rises boost cost of busbar energy. (In Elec. World, vol. 180, Nov. 1, 1973, pp. 40-43)

Soaring fuel costs were by far the biggest factor in boosting total busbar-energy costs during 1972 for the 27 modern steam stations fully reported in the cost survey.# Tables: Summary of key factors reveals pattern of power cost rise over ten-year period; Complete data for 27 modern steam stations show range of total energy costs during 1972. Taxes add mills to energy cost/kw hr when interest rates soar; Unprecedented fuel prices take their toll in busbar energy cost of power.

550 Fuel prices to reshape airlines. (In Astron. & Aeronaut., vol. 13, Sept. 1975, pp. 10-16)

States that two of the nation's major industries and two national policies are turning on a collision course. The administration hopes to make the nation less dependent on foreign oil by raising prices to force conservation and, theoretically, stimulate production. The nation's oil industry applauds that approach. At the same time higher fuel costs would hit the energy-sensitive air transport industry when generally rising costs, soft economy, and static traffic threaten some carriers with bankruptcy and others with an uncertain future.

551 The fuel question, by J. Weiss. (In Air Pilot, June 1979, pp. 5-9; 47-48)

Discusses the outlook on supply and price as still anybody's guess.

552 Fuel state finite, by B. Sweetman. (In Flight Int., vol. 115, Feb. 24, 1979, pp. 548-550, 553-554, 563)

Discusses the continuing rises in the price of fuel, and the prospect of eventual exhaustion of reserves. Looks at the way finite fuel stocks will affect the aircraft industry and some of the technical palliatives.

553 Fuel substitution and price response in UK industry, by E. Bossanyi and I. Stanislaw. (In Energy Economics, vol. 1[2], Apr. 1979, pp. 93+)

554 ICAO gauges future supplies of aviation fuel. (In ITA Bull., no. 1/7, Jan. 1979, pp. 7-9)

Excerpt from an ICAO report. Report indicates that future price trends for aviation fuel are as uncertain as those for the crude oil from which it will continue to be produced.

555 Report of the national scene: fuel prices up in the air, by R.P. Hudock. (In Astron & Aeronaut., vol. 13, Oct. 1975, pp. 11-13, 57)

In petitioning the Civil Aeronautics Board, the DOT, FEA and CWPS estimated the cost of fuel will increase gradually as a result of decontrol by not more than three cents per gallon. The airlines, on the other hand estimate the price rise to be two to three times that amount. The three federal agencies view the proposed emergency procedure to deal with increased fuel costs as limited in duration and self liquidating. "In the long run, higher rate-making load factors--and not higher fares--are preferred means of dealing with increased energy costs." Under the proposed pass-through plan each carrier would recover its own particular increases in fuel costs. The ability of each carrier to recover increased energy costs should be discretionary. Carriers wishing to adjust fares to energy costs should be able to do so in markets of their own choosing.

556 Short range transports to save fuel: a first step design analysis suggests transports designed specifically for ranges out to 500 st. mi.--without reducing any advanced technology--would save 7-10 percent of fuel now burned on those short stages. (In Astron. & Aeronaut., vol. 14, Feb. 1976, pp. 62-64)

States that the shortage of oil and the rapid catastrophic increases in fuel prices necessitate new approaches to aircraft design to reduce the fuel used per passenger mile.

557 Why fuel prices should rise, by R.S. Scorer. (In Coal & Energy Quarterly, no. 20, Spring 1979, pp. 20+)

Statistics

558 Aid to decision making, by R. Martin. (In Pet. Eng., vol. 41, Nov. 1969, pp. PM5-11)

Comprehensive, reliable, and up-to-the-minute statistics on the petroleum industry's overall operation is the foundation supporting day-to-day operations and economic planning of oil company management. Presents a chart of the American Petroleum Institute's statistical department with a brief rundown on the groups, their functions, and their membership.

559 Market data: fuel sales are marking time? (In Can. Chem. Process, vol. 53, Sept. 1969, pp. 54+)

560 Motor fuel consumption data, by Federal Highway Administration. Highway Statistics Division. Washington, D.C.: Public Affairs Office. 1935+
Contains fuel consumption, compiled from state tax records.

V. GAS

Accounting

561 MCD rate for residential service has advantages for gas utilities, by C.A. Larson. (In Pipeline & Gas J., vol. 201, July 1974, pp. 77-79)
Lists and discusses the three main problems which are of overriding concern to management in the natural gas industry: 1, gas supply curtailments; 2, increasing cost of gas supply; and 3, increasing distribution costs for firm service.

562 Operating reserves accounting practices, by R.R. Rice. (In Am. Gas. Assn. Mo., vol. 51, July/Aug. 1969, pp. 23-25)
A study of public utility company practices in accounting for operating reserves indicates a wide variety of individual procedures, tailored to suit the individual environment. Regulatory agencies have shown some interest in operating reserve philosophies and the resultant accounting by gas and electric companies subject to their jurisdiction.# Topics: Self-insurance reserves; Self-insured hazards; Accounting for self-insured losses; Self-insurance reserve provision; Self-insurance reserve balances; Pensions and benefits reserves; Miscellaneous operating reserves.

Costs

563 Alaska gas line at crucial juncture, by F.E. Niering. (In Petroleum Economist, vol. XLVII[3], Mar. 1980, pp. 97-100)
At an overall cost approaching $20 billion, the proposed 4,753-mile pipeline network to condition and transport natural gas from the North Slope of Alaska to the Lower 48 states became an international venture of record proportions.# Headings: Training problems; Gas reserves and pipeline capacity; Pricing Alaskan gas.

564 Economic evaluations of intermediate-BTU gas plants meeting the energy demands, by W.C. Morel. (In Trans. AACE Annual Meeting, 21st, 1977, pp. 372-379)
Presents a summary of the economics of six studies based on four gasification processes. Estimates are based

on January 1976 cost indexes, and the average selling prices of the gas were determined by using discounted cash flow rates of 12, 15 and 20 percent at various coal costs.

565 Economic study of pipeline gas production from coal, by J.P. Henry and B.M. Louks. (In Chem. Tech., Apr. 1971, pp. 238-247)
Partial Topics: Cost of methane from coal; Process heat costs.

566 Economics of developing Canadian Arctic gas, by N.A. Cleland and others. (In J. Pet. Tech., vol. 26, Nov. 1974, pp. 1199-1205)
Examines the costs and potential results of exploring in the area and the well-head prices required to justify the efforts. The well-head prices are translated into prices in the markets that would logically receive such gas.

567 Estimation of the social cost of natural gas, by L.A. Nieves and J.R. Lemon. Richland, Wash.: Battelle Pacific Northwest Laboratories, 1979. 64p. (PNL-3091: EY-76-C-06-1830)
Study determines the extent to which it is possible to develop monetary estimates of the marginal social cost of fuels using natural gas to test a methodology that could be applied to other fuels.

568 Gas plant design can save energy, by C.R. Moncrief. (In Oil & Gas J. Tech., vol. 77[30], July 23, 1979, pp. 55-59)
A more flexible approach to accommodate varying energy costs.

569 Gas turbine cuts operating costs, by J.W. Likens and R.J. Moreland. (In Power Eng., vol. 74, Jan. 1970, pp. 38-39)
Economic and operating experience with 29,000kw gas turbine in the first full year of operation verified Missouri Utilities Co.'s decision to install the equipment.

570 Improving advanced gas cooled reactor economics by reducing initial loading, by A.W. Clarke and D.V. Freek. (In Nuclear Eng., vol. 15, Jan. 1970, pp. 43-46)
Contends that by only partially loading an AGR core for its initial charge while still producing full power output, and then filling the vacant channels as the first refuelling operations, a scheme can be followed in which savings in the order of $2.4/kw are achieved.# Charts: Total fuel costs; Feed fuel uranium costs; Feed fuel fabrication and reprocessing costs.# Tables: Breakdown of replacement fuel costs; Breakdown of costs into total uranium costs and fabrication costs.

571 Koppers-Totzek economics and inflation, by D.M. Mitsak and others. (In Energy Comm., vol. 3[5], 1977, pp. 475-510)

Presents cost data for a large "grass roots" plant designed to gasify Eastern bituminous coal. The effects of coal price, return on equity, and capital structure on the cost of K-T gas are examined.

572 Mini-crew save dollars for utility on distribution construction/service, by A.A. Manchester. (In Pipeline & Gas J., vol. 201, May 1974, pp. 30-32.
Savings made by using two- and three-men gas distribution crews for all types of construction work. Graph shows cost comparisons.

573 New ideas for cutting costs in distribution maintenance, by N.P. Biederman. (In Pipeline & Gas J., vol. 198, May 1971, pp. 44, 46-48)
Reports on three developments helping distribution utilities to cut costs: low noise pavement breaker, soil stabilization, and in-ground checking of pipe condition.

574 Planned costs: a system for achieving results, by R.F. Brunken. (In Pipeline & Gas J., vol. 200, Oct. 1973, pp. 38, 78)

575 Plunging old gas and oil wells ... reduces costs, health hazards, by N.P. Chironis. (In Coal Age, vol. 77, Nov. 1977, pp. 95-98)
Contends that the cost benefits possible with the well-plunging techniques, although secondary to the value of providing safer mining operations, are of considerable importance to mine management.

576 Power recovery cuts energy costs, by S.S. Braun. (In Hydrocarbon Process, vol. 52, May 1973, pp. 81-85)
Shell saves $2½-$3 million a year using expander turbines on cat cracker flue gas stream, plus $2 million a year with hydraulic turbines for liquid streams. Explains how it's done.

577 A preliminary evaluation of the cost of natural gas deregulation. [Washington]: Federal Power Commission, Inter-Agency Task Force, 1975. 1v. (variously paged)
An evaluation of the cost of alternative proposals for deregulating the wellhead price of natural gas sold in interstate commerce.

578 Rising costs, by D. Hale. (In Pipeline & Gas J., vol. 197, May 1970, pp. 43-47)
Survey examines the cost problems of the gas distribution utility industry and solutions companies are using in the continuing battle against rising costs.# Tables: Materials and equipment cost index; How costs have risen in gas distribution; January 1970 gas distribution installed cost index.

579 Role of gas and steam turbines to reduce industrial plant energy costs, by W.B. Wilson and W.J. Hefner. (In Combustion, vol. 45, Nov. 1973, pp. 32-41)
Partial Charts: Fuel changeable to power with 850 PSIG-825F steam turbines; Fuel changeable to power with 1450 PSIG 950 steam turbines; Gas turbine fuel treatment plant investment costs.# Partial Tables: Heavy fuel treatment operating costs; Average total maintenance cost MS-5001 gas turbine.

580 Some analysis and projections on liquid and gas production from oil shale, by G.E. Klinzing, S.H. Chiang and J.T. Cobb. (In Energy Comm., vol. 6[1], 1980, pp. 17-39)
Considers the cost of producing oil from shale.# Headings: Cost comparison; Production costs from individual process reports; Critical cost comparison.

581 Well completed report: economics dictates six-pay mini-completion, by G.O. Ires. (In Pet. Eng., vol. 41, Aug. 1969, pp. 54-55)
Discusses the high rig costs and the need to institute an economic development program for the multipay West Part Isabel gas field for maximum production.

Finance

582 A critical analysis of the technology and economics for the production of liquid and gaseous fuels from waste, by S.H. Chiang, J.T. Cobb and G.E. Klinzing. (In Energy Comm., vol. 5[1], 1979, pp. 31-73)
Utility and private financing fuel gas range in cost from $3/10^6 BTUs to $5/10^6 BTUs. These costs of gas from waste are comparable to the projected costs of gas from coal gasification processes. Production costs of liquids are seen to be more expensive than the gas costs with a less developed technological basis.# Partial Table: Cost of gas from wastes.

583 Financial appraisal of the gas industry. (In Am. Gas Assn. Mo., vol. 56, Nov. 1974, pp. 22-24)
An excerpt from a speech given by H. Fraser, at the AGA Financial Seminar for gas companies in Boca Raton.

584 Financial statistics of the major privately owned utilities in New York State: electric gas.... New York: Department of Public Service, Accounting Systems Section. 1926+ Annual.

585 Financing the capital needs of the natural gas industry, by R.H. Blanchard. (In Am. Gas Assn. Mo., vol. 56, Jan. 1974, pp. 25-26, 37)

Paper presented at the Financial and Administrative Section luncheon, AGA Convention, emphasized that the gas industry in the period of 1973 to 1985 is going to have needs for capital in a financial market where capital is going to be a scarce and precious commodity. The days of financing the largest part of capital expenditures from internal cash flow are gone, it states.

586 Financing the gas industry's burgeoning capital needs, by R.H. Blanchard. (In Energy Pipelines & Systems, vol. 1, Feb. 1974, pp. 46-48)

587 Gas company financing: a perspective on the problem, by B.A. Wigmore. (In Am. Gas Assn. Mo., vol. 54, Dec. 1972, pp. 29-31, 36)

Based on the theme that gas distribution companies, except for a small few among them, pay serious financial penalties because they are too small, and mergers are the solution.

588 Gas from the UK continental shelf. London: Department of Energy. 1976. 7p. (Fact sheet 3)

Describes progress in the development of natural gas from the UK sector. Some of the areas covered include: reserves and supplies; production; benefits to the balance of payments; government revenue; and depletion policy.

589 Gas industry capital requirements for energy independence. (In Am. Gas Assn. Mo., vol. 56, Oct. 1974, pp. 13-16)

"Consideration should be given to regulatory practices that would involve adding the value of leased property to the rate base, thereby supporting earnings in the same way as do property additions financed in a more conventional manner." Excerpts from the testimony of W.R. Thomas at the New York hearing sponsored by the Federal Energy Administration.

590 Investment community views on gas industry securities, by R.H. Healy. (In Am. Gas Assn. Mo., vol. 55, Dec. 1973, pp. 16-18)

Describes how the institutional investor evaluates the natural gas industry--a survey of the entire spectrum of equity investment opportunities.

591 1974: an increase in revenues, customers. (In Am. Gas Assn. Mo., vol. 57, Feb. 1975, pp. 14-15)

Growth rate continues to fall; future gas sales to come from sources other than conventional production. Includes table showing year-end estimate, total gas utility industry, customers, sales and revenues--preliminary 1974 compared with 1973.

592 Norwegian gas. (In Noroil, vol. 7, Nov. 1980, pp. 20-21, 23)

It's money in the bank, and only European incentives can persuade Norway to make a withdrawal.

593 Paying and repaying our way, by W.G. Jewers. (In Gas World, vol. 183, Jan. 1978, pp. 13-15, 19)
During the past year British Gas Corporation consolidated its profit margin and laid a solid foundation on which to build up its financial strength into the future....

594 Problems of project financing in the gas industry. (In Am. Gas Assn. Mo., vol. 56, Sept. 1974, pp. 10-13)
Highlights of David O. Beim's paper at the 1974 AGA Executive Conference in which he discussed in depth the potential of project financing as a means of obtaining the capital which will be required by the gas industry for such major programs as coal gasification, synthetic gas plants, and transport storage and distribution of LNG.

595 Profit picture brightens. (In Pet. Eng., vol. 43, Mar. 1971, pp. EM5-10)
Presents fourth quarter financial reports for oil and gas companies.# Tables: 1, Oil company financial reports; 2, Natural gas company reports.

596 Projection and conjecture on growth, by L.W. Fish. (In Am. Gas Assn. Mo., vol. 54, July/Aug. 1972, p. 27)
State of gas utilities--common stocks are selling at extraordinarily low multiples: bonds are being down rated, and investors are worried about gas supply shortages and implications on future sales.

597 Prospects for world natural gas: a Bank of Scotland Information Service. Edinburgh: Bank of Scotland, Oil Department, [1979]. 24p. illus.
Partial Contents: The international natural gas trade--natural gas prices: the key to market growth; financing natural gas development.

598 Securities markets: will the financial community be able to provide business access to capital at reasonable cost? by G.L. Levy. (In Am. Gas Assn. Mo., vol. 56, Oct. 1974, pp. 26-28)
Abstract of a paper deliverd at the AGA Financial Seminar, Boca Raton, Florida, in which the author discussed the securities markets' structure and developing trends as they affect the opportunities for gas industry financing in the near and intermediate future.

599 $3.5 billion budgeted by companies to assure a healthy gas industry, by K. Kridner. (In Energy Pipelines & Systems, vol. 1, May 1974, pp. 39-46)

Management

600 Elasticity measurements and value of service, by D. Parson. (In Pipeline & Gas J., May 1970, pp. 92, 94, 96)
Discusses what the customers want and what they'll pay for gas.

601 The exploitation of Norwegian oil and gas, by H. Ager-Hanssen. (In Energy Policy, vol. 8[2], 1980, pp. 153-164)
Topical Headings: Regulatory policies; The taxation policy; Current investment and the production profile; The impact on Norwegian national income.

602 Gas marketing during supply shortages, by D. Parson. (In Pipeline & Gas J., vol. 197, June 1970, pp. 49-50, 52)
Puts forward new management concepts.

603 Gas men's round table. (In Am. Gas Assn. Mo., vol. 53, July/Aug. 1971, pp. 12-14)
A monthly meeting of industry and government executives provides a forum for issues of timely interest to the gas industry.

604 Major energy market, how one dual utility approaches it, by R.D. Buckman. (In Am. Gas Assn. Mo., vol. 54, Sept. 1972, pp. 4-7)
Paper presented at the 1972 Pacific Coast Gas Association Marketing Conference.

605 Management: who needs it? by J. L. Hayes. (In Am. Gas Assn. Mo., vol. 53, Dec. 1971, p. 31)
Excerpts from address presented at the 53rd Annual AGA Convention, Boston, Massachusetts.

606 1974 management tools, by R.H. Murray. (In Am. Gas Assn. Mo., vol. 56, Jan. 1974, pp. 33-37)
A paper presented at the Marketing Section meeting, AGA Convention, Oct. 16, 1974. It discusses the Federal Power Commission's National Gas Survey, sketching some of the provocative findings. All task forces were charged with presenting and analyzing historical data and forecasting for the years 1974-80, 1980-85, and 1985-90. All dollar forecasts are made in current dollars reflecting an inflation rate of 4 percent per annum for 1970 through 1975, 3½ percent from 1976 through '80, and 3 percent from 1980 through 1990.

607 The Rock: a training ground for young managers. (In Am. Gas Assn. Mo., vol. 53, Nov. 1971, pp. 8-11)
Expresses the need for active development of good managers in the Northern Illinois Gas Company.

608 Selective marketing. (In Am. Gas Assn. Mo., vol. 52, July/ Aug. 1970, pp. 14-15)
How to market in the face of a tight gas supply occupied a good proportion of the attention of industrial and commercial gas salesmen who met in Minneapolis, May 5-7, for the annual AGA 1-C Gas Sales Conference.

609 Starting a 48-plant EVOP program, by H. Grekel and L. E. Childers. (In Chem. Eng. Prog., vol. 69, Aug. 1973, pp. 93-97)
Gives tips on how the evolutionary operation technique was used to improve efficiency and reduce cost in a large number of natural gas processing plants.

610 Uncommon plant problems of common ownership and their common solutions, by H. J. Bretzger. (In Am. Gas Assn. Mo., vol. 52, Apr. 1970, pp. 25-28)
Defines common problems which include: Establishment of a system of cost record keeping; Definition of shared and unshared costs; Resolution of depreciation considerations; Handling of cash; and, Classification of legal and administrative matters.

Pricing

611 British Gas Corporation: gas prices and allied charges, by the Price Commission. London: HMSO, 1979. ix, 117p. (1978/79 HC-165)

612 The coming rise in gas prices. (In Petroleum Economist, vol. XLVI[2], Feb. 1979, pp. 46-47)
Gas will rise in price on the world market as growing pressure of demand drives home the lesson that this high grade fuel is being sold in many countries at well below the long-term supply prices. # Headings: Low selling prices; Economic consequences; Cost considerations; Facing reality.

613 Decontrol of natural gas prices hoped for, by E. V. Anderson. (In Chem. Eng. N., vol. 53, Aug. 25, 1975, pp. 12-16)
Reveals that about 80 percent of the natural gas used as fuel by the petrochemical industry comes from intrastate sources (chiefly Texas and Louisiana). This gas isn't price-controlled, but already is responsive to market demand. Thus the burden of decontrolled gas prices will fall upon only the 20 percent of the petrochemical industry's requirements that come from the interstate pipeline.

614 Factors affecting estimates of natural gas reserves, by R. D. Stanwood. (In Am. Gas. Assn. Mo., vol. 57, May 1975, pp. 28-29)
Abstract of paper presented at the American Petroleum

Institute Reserves Seminar, April 1975, Washington. # Topical Headings: Economic facts; Economies affect reserves; Wide ranging gas prices.

615 FERC implements incremental gas pricing plan. (In Oil & Gas J., vol. 77[39], Sept. 24, 1979, pp. 84-85)
Comment on a compromise incremental pricing plan designed to discourage US industrial users of natural gas from switching to fuel oil, as implemented by the Federal Energy Regulatory Commission.

616 Gas: its price and future role in the energy scene, by O. Al-Rawi. (In OPEC Bull., Aug. 1980, pp. 20-28)
Contents: Prices of LNG and natural gas; LPG pricing--linking the price of LPG to end uses; Price of LPG as a petrochemical feedstock competing with naphtha and gas oil (50/50 percent). Tables: 1, World natural gas and LNG trade; 2, Evolution of gas prices in comparison with crude oil and refined products in Europe; 3, Evolution of LPG prices in relation to the marker crude; 4, Share of LPG in the major markets in 1980.

617 Gas purchase contract: who should participate? by D. Parson. (In Pipeline & Gas J., vol. 199, June 1972, pp. 96, 98, 101)
An approach to gas supply/price problems that recognizes that pipelines are primarily in the transportation business. Considers implications of higher cost supplies.

618 A marginalist approach to pricing US natural gas, by R. Smiley. (In Energy Econ., vol. 2[3], July 1980, pp. 172-178)
The current method of retailing pricing for natural gas in the USA introduces serious distortions in the allocation of natural gas. The cost to the consumer is, in some instances, less than half of the marginal cost of reducing the gas. Paper presents an alternative to the gas pricing system in use in nearly every utility in the USA. # Headings: Marginal costs for a typical company; pricing and economic efficiency; Second best pricing; The problem of excess revenue.

619 Market forces mulled for gas price, curtailment. (In Oil & Gas J., vol. 77[29], July 1979, p. 42)
Comment on the use of market prices rather than regulation as the basis for incremental pricing and curtailment policies for natural gas.

620 The natural gas shortage and the Congress, by P. E. Starratt. Washington: American Enterprise Institute for Public Research, 1974. [6], 68p.
Partial Contents: The debate over field price regulation; Effects of field price regulation; Competition in field markets.

621 OPEC gas and its price, by F. Azarnia and B. Andrade. (In OPEC Bull., Sept. 1980, pp. 21-26)
Tables: Evolution of gas in comparison with crude oil and refined products in Europe; 2, Evolution of LPG prices in relation to the market crude.

622 Operators scramble to trap deep gas in South Louisiana, by L.J. Pankonien. (In World Oil, vol. 189[4], Sept. 1979, pp. 55-62)
With gas prices nearing the $3Mcf mark, operators have found an attractive gamble in 17,000 to 20,000 foot gas of the Tuscaloosa trend. Production operations remain costly operations.

623 Price controls and the natural gas shortage, by P.W. MacAvoy and R.S. Pindyck. Washington: American Enterprise Institute for Policy Research, 1975.
Contents: Is there a natural gas shortage?; The Federal Government's policy toward natural gas; What can be done? --maintain control, price increases, strengthen control, price freezes; Industry response; The effects of gas policy changes on producers, consumers and others. # Appendix: The econometric model of natural gas.

624 The price of natural gas, by R. Fort. (In Revue de l'Energie, Aug./Sept. 1979, pp. 667+)

625 Regulating price of natural gas, by R. Moddy. (In Am. Gas Assn. Mo., vol. 55, Nov. 1973, pp. 28-30)
Partial Topical Headings: Opening the doors to deregulation; Consumerism vs. pricing; Balancing supply and demand. # Paper presented at the AGA on wellhead price regulation of natural gas after 19 years.

626 Special report on world natural gas pricing, by J. Segal and F.E. Niering. (In Petroleum Economist, vol. XLVII[9], Sept. 1980, pp. 373-379)
Schemes for deciding prices or linking gas prices to those of alternative fuels or non-fuel indices increased in number and variety while effective cohesion in policy-making continued to elude the major gas exporting countries. # Tables: 1, World natural gas import prices, 1975-1979; 2, Gross calorific values of crude oil and products; 3, Projected OPEC gas export prices at crude parity....

627 TERA looks at tomorrow's gas prices, by D.O. Dawson and J.R. Sharko. (In Am. Gas Assn. Mo., vol. 55, July/Aug. 1973, pp. 26-28)
Tables: New and average wellhead prices-base case ($Mcf); Supplemental sources of gas; 1980 regional average City Gate prices of supplementals ($Mcf); Usage per customer and cost per customer for the Middle Atlantic region, 1920, 1980-1990; Percentage changes in fuel prices for the

period 1970-1980 ...; Projected retail fuel prices for the Middle Atlantic region for 1980.

Rates

628 Natural gas still a clean bargain, by J.M. Beall. (In Am. Gas Assn. Mo., vol. 54, June 1972, pp. 20-22)
Natural gas, the cleanest burning of the major fuels, will take a slightly larger bite from the consumer's pocketbook in future years. Yet it will still be among the householder's biggest bargains when compared to other items in the rising cost of living.

629 Optimizing continuous flow gas lift wells, by W.E. Simmons. (In Pet. Eng., vol. 44, Aug. 1972, pp. 46-48)
Chart: Oil production rate. # Table: Incremental income vs. gas injection rate for hypothetical well.

630 Saving money through interruptible gas rates, by F.D. O'Neill. (In Chem. Eng., vol. 79, Feb. 7, 1972, pp. 96, 98, 100, 102)
By agreeing to accept occasional cut-offs in natural gas, and by being billed at the "interruptable gas rate, rather than the more expensive" firm rate, significant amount of money can be saved.

Statistics

631 Consumer income vs. consumer energy consumption, by J.H. Climer. (In Am. Gas Assn. Mo., vol. 56, May 1974, pp. 28-29, 31)
Calculating the relationship of income to consumption for the residential customer. # Topics: Sample data gas service area--gas residential space heating; Residential gas space heating customers distribution of monthly consumption; Residential gas non-space heating customers distribution of monthly consumption.

632 Emergence of US gas utilities as a factor in world petroleum economics, by E.F. Hardy. (In Am. Gas Assn. Mo., vol. 56, May 1974, pp. 7-11)
Tables include: Natural gas supply and demand to lower 48, 1970-1990; Canadian gas imports, 1970-1990; Canadian drilling, 1970-1972; LNG imports to US, 1970-1990. # Indicates that gas production in the lower 48 states could be increased steadily through the 1980s if stimulated by reasonable new gas prices.

633 Factors controlling gas supply and demand, by J.C. Keaton. (In J. Pet. Tech., vol. 23, Jan. 1971, pp. 41-44)
If the gas industry were permitted to operate under the

economic laws of supply and demand, the wellhead price per Mcf of gas would conceivably jump by as much as 15 cents. Chart: Natural gas consumption since 1948, with projection through 1990.

634 Gas heat goes up. (In Am. Gas Assn. Mo., vol. 53, Nov. 1971, pp. 30-31)
Tables: Number of house heating customers, 1961-1969-1970; Housing units heated by utility gas, 1961-1969-1970. # Anticipated additional house heating customers of gas utilities by state, 1971-1973.

635 Gas industry continues long-term growth in 1969. (In Am. Gas Assn. Mo., vol. 52, Jan. 1970, pp. 14+)
Contains figures for total gas utility industry customers, sales and revenues preliminary 1969 compared with 1968.

636 Gas industry 1971 year end report. (In Am. Gas Assn. Mo., vol. 54, Feb. 1972, pp. 18-19)
Includes table showing year-end estimate, total gas utility industry customers, sales and revenues.

637 Gas utility and pipeline industry projections, 1969-1973, 1975, 1980, 1985 and 1990. (In Am Gas Assn. Mo., vol. 51, Oct. 1969, suppl. i-viii).
US Department of Statistics presents annual long-term projections of gas utility and pipeline industry operating statistics including forecasts of economic indicators used in the development of the gas industry projections.

638 Gas utility industry projections to 1990. (In Am. Gas Assn. Mo., vol. 55, June 1973, pp. 21-36)
Sales of the gas utility industry are projected in order to reflect data related to these industry operations rather than the total gas demand in the United States. # Tables include: Total gas utility industry sales by region, annual indexes of general economic indicators, and US gas utility industry sales.

639 How 100 energy companies fared, by E. Adams. (In Pet. Eng., vol. 44, May 1972, pp. EM11-18)
Tables: 1, Financial statistics of US petroleum companies; 2, Financial statistics of natural gas companies; 3, Drilling results by companies; 4, Natural gas and AGL producers; 5, Liquid hydrocarbon producers among US companies; 6, Refining operations by companies; 7, Petrochemical sales; 8, Natural gas pipelines; 9, Refined petroleum product sales; 10, Free expenditures for facilities and reserves.

640 Natural gas at any price, by B.M. Grunnell. (In J. Pet. Tech., vol. 22, Aug. 1970, pp. 937-940)
Topics: Requirements for gas; Rates of reserves to production; Available gas supplies; Price. # Tables: Forecasts

of natural gas requirements in the contiguous United States. #
Charts: Gas consumption in the United States, historic and forecast (DeGolyer and MacNaughton forecast); Gas reserves and reserve production ration for US; Correlation, prices vs. additions to non-associated gas reserves in US.

641 1972 reserves and production report. (In Am. Gas Assn. Mo., vol. 55, May 1973, pp. 11-12)
Includes tables showing summary of annual estimates of natural gas reserves in the United States for period December 31, 1945 to December 31, 1972; and 1972 remaining marketable natural gas reserves in Canada.

642 Production and reserves report. (In Am. Gas Assn. Mo., vol. 53, Apr. 1971, pp. 7-8)
Table includes 1970 estimates of natural gas reserves in Canada--Canadian Petroleum Association.

643 24th Annual gas industry house heating survey. (In Am. Gas Assn. Mo., vol. 55, Feb. 1973, pp. 16-18)
Tables shown include: Residential and house heating consumers of gas utilities by state, December 1931, 1971, and anticipated additional house heating customers of gas utilities by state, 1972-1974.

VI. NUCLEAR ENERGY

Costs

644 Bomb, unlimited fuel, cheap power and little pollution, by C. Beatson. (In Engineer, vol. 236, May 17, 1973, pp. 48-49, 51)
Discusses a utopian prospect of harnessing thermonuclear fusion for low-cost power generation which is being explored by British researchers.

645 Boon to society: the LMFBR, by R.J. Creagan. (In Mech. Eng., vol. 95, Feb. 1973, pp. 12-16; Discussion, vol. 95, July 1973, pp. 62/63)
Gives a rundown based on the Westinghouse design prepared during the project definition phase of the liquid-metal fast breeder reactor (LMFBR) demonstration program of the US AEC. # Tables: 1, National investments in LMFBR; 2, LMFBR projects; 3, US electric utility power statistics; 4, Fuel cycle cost.

646 Breeder reactor costs ballooning. (In Chem. Eng. N., vol. 53, May 12, 1975, pp. 7-8)
Discusses a study by the General Accounting Office, in which it was claimed that the current fast-breeder reactor program eventually could cost the federal government an additional $1.7 billion in subsidies before it becomes economically competitive with other forms of energy.

647 Capital cost escalation and the choice of power stations, by K.R. Shaw. (In Energy Policy, vol. 7[4], Dec. 1979, pp. 321-328)
Examines the increase of AGR construction costs in the UK and LWR costs in the USA and Germany, and shows that with expanding nuclear power plant material requirements in the last decade, costs have risen considerably faster than inflation. # Topical Headings: Inflation; Capital cost growth; Capital cost estimates. # Charts: 1, Net UKAEA expenditure on nuclear R&D as a percentage of the value of electricity sales; 2, US nuclear construction cost estimates for a 1000 MWe LWR; 3, German nuclear construction cost estimates for a 1000 MWe LWR; 4, Cost comparison of 2000 MW coal and nulcear stations, showing effect of future

growth rates of coal costs and nuclear building costs. # Appendix Tables: A1, Cost and construction durations for major US light water reactors, 1970-1977; A2, Effect of capital cost growth rate on the relative economics of nuclear (AGR) and coal stations (2000 MW); A3, Capital costs of AGR and coal generation; A4, Coal cost profiles.

648 Capital cost of calculations for future power plants. (In Power Eng., vol. 77, Jan. 1973, pp. 61-65)

A rational, consistent and rapid way to calculate future plant costs developed by AEC. These procedures cope with the steep upward spiral in costs and the constant impacts of new laws and regulations. # Topics: Computerized cost estimating; Base cost models; Typical plant estimate; Cost of 1980 plants; Capital plant sensitivity study; Capital costs vs. location; Projected future capital costs. # Tables: Capital estimates for 1000-MWE steam electric power plants for operation in 1980 at Middletown; Sensitivity of power plant capital costs to various parameters (four tables); Projected future capital costs of nuclear and fossil power plants ...; Unit capital costs....

649 Charts aid study of fuel-cycle cost changes from pending shift in diffusion plant ownership. (In Elec. World, vol. 172, Nov. 1969, pp. 34-35)

Although details of facility costs and makeup of separate work costs have been published, they do not equate variations in separative work parameters into the resultant fuel cost change that an operating utility would experience. To simplify their own evaluations of the future fuel cycle costs, Babcock & Wilcox engineers have prepared a set of easy-to-use graphs. They permit a rapid determination of the change in fuel cost that results from various combinations of diffusion plant parameters.

650 Choosing your next plant? by K.W. Hamming. (In Elec. World, vol. 173, Mar. 16, 1970, pp. 32-34)

Examines some of the changing evaluation factors. # Tables: 1, Direct plant costs for a 900-Mw unit distribution among major categories approximately as shown. Costs for the coal unit are based on the total cost for the nuclear unit (without the cooling tower) taken as 100 percent; 2, Total delivered costs for two principal types of fuels show how procurement costs and shipping charges contribute to the total cost; 3, Capital requirements for an equilibrium fuel batch for a 1,000 MWe light water reactor illustrate the action of a typical payment schedule; 4, Cumulative capital requirement of a fuel cycle for a 1,000 MWe light water reactor shows peaks reflecting partial fuelling which begins in the sixth year.

651 A comparison of the capital costs of light water reactor and liquid metal fast breeder reactor power plants: prepared

for the Arms Control and Disarmament Agency by W.E. Mooz and S. Siegel. Santa Monica, Cal.: Rand, 1979. ix, 65p. (Rand Corp. R-2441-ACDA)

652 Core management and fuel costs: abstract, by A.J. Bauks and others. (In Combustion, vol. 41, Aug. 1969, pp. 32-33)

Abstract of a paper presented at the American Power Conference, 1969.

653 Developing world's need for nuclear, by D. Kovan. (In Nuclear Eng., vol. 20, Jan. 1975, pp. 38-42)

A review of an assessment by the International Atomic Energy Agency on the economics of nuclear power in developing countries. # Tables: Breakeven plant factors and oil prices for market survey countries; 2, Market for nuclear plants by size of unit (1981-1990); 3, Influence of oil price, plant factor and interest rate on market for nuclear plants, 1981-1990, GWe; 4, Breakeven of oil prices as a function of plant size, plant factor and interest rate; 5, Economic parameters used in sensitivity studies; 6, Sensitivity of the nuclear power market to changes in economic parameters; 7, Capital cost data, $kW (1974).

654 Developments in nuclear power economics, Jan. 1968-Dec. 1969, by P. Sporn. (In Combustion, vol. 41, June 1970, pp. 14-20)

The light water moderated reactor, which two years ago offered potentials for nuclear power generation comparative with fossil fuel at 22¢ to 24.8¢ per million Btu, has today lost position where it is competitive at 28¢ to 29.5¢ per Btu fossil fuel cost.

655 Dry cooling affects more than costs, by K.A. Oleson and others. (In Elec. World, vol. 178, July 1, 1972, pp. 50-54)

Application of dry cooling towers requires a determination of the overall economics and operating requirements of the turbine, condenser, and cooling system. Paper is based on an interdisciplinary study conducted by Westinghouse's Large Steam Turbine Division, Heat Transfer Division, and Power Generation Systems. # Partial Tables: Comparison of generation costs; 1978 cost evaluation for dry cooling (1,132 Mw nuclear plant).

656 Economic evaluations of nuclear power plants using Monte Carlo simulation: EVANC (Evaluation of Nuclear Costs), by J.B. Hoge. (In Power Eng., vol. 78, Dec. 1974, pp. 36-38)

Comprehensive and rapid economic comparisons of up to 25 nuclear power plants are possible by use of new computer code. The code was developed by Babcock & Wilcox. It is called EVANC (Evaluation of Nuclear Costs). # Partial

Tables: 1, EVANC evaluation equations; 2, System and project parameters; 3, NSS fuel costs, heat rate, and terms of payment; 4, Total plant cost--millions of dollars; 5, Plant economic evaluation.

657 Estimating average nuclear fuel costs, by J. F. Bader and others. (In Power Eng., vol. 73, Dec. 1969, pp. 50-53)
System for expediting estimates of average nuclear fuel costs for each fuel region of each core loading. Includes capital investment diagram for one annual reload region. # Tables: Nuclear fuel cost estimate--economic assumptions; Nuclear fuel cost estimate.

658 Evaluating nuclear fuel cycles: decision analysis and probability assessments, by A. S. Manne and R. G. Richels. (In Energy Policy, vol. 8[1], Mar. 1980, pp. 3-16)
Partial Charts: 1, Levelized busbar cost assumptions; 2, The lower row of figures shows capital costs; 3, Present value of costs and benefits of alternative nuclear fuel cycles; 4, Cost summary for research, development and demonstration program. # Partial Tables: Cost assumptions for economic comparison of US base load electricity plants (1975 general price level); 3, Expected money value of the economic benefits....

659 The fast breeder reactor: need? cost? risk? edited by C. Sweet. New York: Macmillan, 1980. 232p.

660 Fast breeder: schedule lengthens cost estimates, by F. C. Olds. (In Power Eng., vol. 76, June 1972, pp. 33-35)
In 1967 a demonstration LMFBR was proposed for 1975 operation. Its cost was estimated at about $100 million. It was never built. Today's demo has a post-1980 operating date and cost estimates range from $390 to $675 million. # Tables: 1, Capital costs of first-of-a-kind LMFBR's on a 1973 basis; 2, Total investment liability for a 600 MWe LMFBR demo unit.

661 How to save on nuclear insurance. (In Elec. World, vol. 172, Aug. 4, 1969, pp. 113-116)
A management newsletter which contains a timetable of risk management considerations for proposed nuclear plants, 1970-1975.

662 The need for cost benefit perspective in nuclear regulatory policy, by E. P. O'Donnell. (In Nuclear Eng., vol. 24[291], Oct. 1979, pp. 47-51)
Contends that spending by government agencies on measures to improve health and safety is not cost effective. # Tables: 1, Cost-benefit ratios for various health and safety protective measures; 2, Costs incurred in forced shutdown of four nuclear plants for 75 days; 3, Improved care programs available for equivalent costs incurred in forced shutdown of four nuclear plants for 75 days.

663 Nuclear costs; the future looks better, by J. Wright. (In Elec. World, vol. 173, Feb. 9, 1970, pp. 26-29)
Paper based on examining such construction-cost factors as engineering, labor, construction scheduling, interest and escalation.# Partial Tables: Plant cost comparison for 1969 operation; Annual cost escalation--for US construction; Field labor costs; Future plant costs--nuclear plants, coal plants; Future cost trends for the PWR.

664 Nuclear power and the money managers. (In Elec. World, vol. 179, Apr. 15, 1973, p. 13)
A brief report on a meeting of a panel of legal and financial experts brought about in Chicago by the American Bar Association and the American Law Institute.

665 Nuclear power costs in the UK, by C. Sweet. (In Energy Policy, vol. 6[2], June 1978, pp. 107-118)
Assessment of the true costs of nuclear power has been oversimplified and the supposed benefits have yet to be demonstrated. Calculation of the true costs has been obscured by the use of mistaken assumptions and methods.# Partial Headings: GECB generating costs; Capital costs; Fuel cycle costs; Net benefits.# Partial Tables: Estimated generating costs at nuclear stations at 75 percent load factor; Cost of nuclear stations; Nuclear fuel cycle costs; Estimated nuclear costs in the UK, 1975 and 1980 (p/kWh).

666 Nuclear power engineering: analysis of trends in policy and technology--fuel cycle insurance, by F.C. Olds. (In Power Eng., vol. 74, Sept. 1970, pp. 16+)
Shows possible types of policies and sample costs for typical periods of coverage.

667 Nuclear power for the underdeveloped? (In Elec. World, vol. 173, Jan. 12, 1970, pp. 24-26)
Although growing rapidly, power demand in the developing nations is usually not large enough to justify nuclear plants in size ranges that are economic.# Tables: Plant costs--are higher for nuclear power; Annual fixed charges --highlight penalty for nuclear's higher investment; Fuel cost advantage; Unit generating costs.

668 Nuclear project budgets rise; hydro fares badly. (In Elec. World, vol. 179, Feb. 15, 1973, p. 23)
Energy and power programs in the 1974 federal budget reflect the general tone of the administration's fiscal program--deep and wide cuts. In the energy area, many of these dealt with water-related projects, while projects as the fast reactor breeder and nuclear fusion continued to get solid support.# Tables: What power agencies plan to spend --Atomic Energy Commission; Interior Department; Tennessee Valley Authority; Rural Electrification Administration; Federal Power Commission; Corps of Engineers; Environmental Protection Agency.

669 Operating cost for nuclear and fossil-fire plants: data sheet, by N. Peach. (In Power, vol. 116, Dec. 1972, p. 98)
Presents a chart which shows that operating cost of a power plant consists of the total fuel cost plus operating and maintenance (O&M) expenses.

670 Problems and prospects for nuclear power, by W.K. Davis. (In Chem. Eng. Prog., vol. 66, Feb. 1970, pp. 15-21; Discussion, by R.B. Olney, vol. 66, May 1970, p. 12)
Contends that the prospects for nuclear power use in the US and abroad are brighter than they have ever been. The problems of costs, schedules, operating difficulties, etc., while real and frustrating, can be overcome and are not fundamental to nuclear power.

671 Realizing the potential of the breeder, by L.A. Feathers. (In Combustion, vol. 45, Aug. 1973, pp. 31-35)
Paper presented before the American Power Conference, Chicago, May 1973.# Partial Topics: Economic considerations; Fuel cycle costs; Capital costs.# Partial Tables: Estimated early Liquid Metal Fast Breeder Reactor (LMFBR) fuel cycle cost ...; Cost effect of various levels of LMFBR design technology.

672 Social cost: benefit analysis and nuclear features, by D.W. Pearce. (In Energy Econ., vol. 1[2], Apr. 1979, pp. 66+)

673 Social costs and benefits of nuclear futures, by D.W. Pearce. Aberdeen: University, Department of Political Economy, 1979. 32p.
A discussion paper.

674 Technical and economic aspects of uranium enrichment in Europe. (In Nuclear Eng. Int., vol. 14, July 1969, pp. 580-583)
A report on a one-day international symposium organized by the Netherlands Atoomforum at Utrecht, Holland on May 30, 1969.# Partial Tables; Cost of separative work by UC and gas diffusion methods; Approximate specific work separative work cost envisaged for the 700,000kg SWU/y demonstration nozzle separation enrichment plant; Comparison of capital costs established for three types of separation plants.

675 US defines fuel shortage costs, by G. Greenhalgh. (In Energy Int., vol. 16[2], Feb. 1979, pp. 22-23)
Table shows cost breakdown ($/kg) of heavy metal.

676 Will nuclear power pay off? by T.N. Marsham. (In Electronics & Power, vol. 17, Feb. 1971, pp. 57-63; Discussion, by A.E. Lott, July 1971, p. 278)
Partial Charts: 1, Electricity generating capacity trends, 1970-2000; Mean station load factors, 1970-2000; Power station installation rates, 1970-2000; Expenditure comparison

between nuclear and fossil power stations based on a 2000 MW station at 75 percent load factor; Comparison of foreign exchange requirements between nuclear and fossil programs.# Table: Approximate annual expenditure on fast reactor nuclear power stations in the period 1980-2000 assuming 50,000 MW.

Finance

677 Designing for cycle boiler operation. (In Power, vol. 115, Feb. 1971, pp. 55-56)
General acceptance of large nuclear and fossil units for base-load operation calls for low-capital-cost boilers to satisfy medium load-factor demand.

678 Effect of inflation and recession on nuclear project financing in the USA, by J.S. Katzin. (In Nuclear Eng., vol. 20, Mar. 1975, pp. 184-186)
Partial Headings: Excessive cost of licensing; Utility costs and stock values; Effects on nuclear project financing techniques.

679 Financing nuclear power for Germany, by T. Roser. (In Nuclear Eng., vol. 20, Mar. 1975, pp. 177-178)
Utilities' ability to provide adequate funds from their own resources depends on whether they can change proper economic prices for electricity.# Headings: Capital requirements; Financial resources of utilities; External sources of finance.# Table: Estimates of German investment requirements.

680 Financing nuclear power in Japan, by I. Suetsuna. (In Nuclear Eng., vol. 20, Mar. 1975, pp. 179-182)
Governmental financing is being carried out to a considerable extent at all stages from research and development through full-scale demonstration to commercial series production of large-scale nuclear power stations for the nine major electric power companies.# Table: Finances raised by the nine electric power companies, annual since 1966 as of December 20, 1974, in billion Yen.

681 Financing nuclear power projects: changes in the way Japanese utilities raise their funds, by I. Suetsuna. (In Nuclear Eng. Int., vol. 24[285], May 1979, pp. 65-67)
Japan is heavily committed to the use of nuclear power but with the long lead times large amounts of capital are at present being invested without financial return. This is having a detrimental influence on the financial structure of the utilities, particularly in regard to self-financing.# Headings: Depreciation methods: Bond issues.# Tables: 1, Equipment investment, internal funds and depreciation--nine

electric power companies; 2, Fixed assets and depreciation --nine electric power companies

682 Financing nuclear power projects: delays help German utilities maintain self-financing ratios, by G. Radthe. (In Nuclear Eng. Int., vol. 24[285], May 1979, pp. 64-65)
Estimates of electricity consumption have been substantially reduced and nuclear plant is now expected to be 22 percent of total generating capacity of 1985 instead of the earlier forecast of 36 percent. The decline in the ordering of new plant has benefited the financial position of the electricity utilities and the expected fall in self-financing ratios has not occurred.

683 Financing nuclear power projects: meeting the special requirements of the fuel cycle industry, by N.C. McKenzie. (In Nuclear Eng. Int., vol. 24[285], May 1979, pp. 59-62)
The nuclear cycle industry is generally thriving, despite the slow-down in nuclear power plant construction. Turnover is expected to grow from the present $4.7 billion to $12 billion in 1985 and nearly $16 billion in 1988. Special nuclear fuel finance companies have been formed to help satisfy their financing needs.# Headings: Requirement for funds; Security for loans; Special types of contract.# Tables: 1, Turnover in the nuclear fuel business; 2, Types of financial instruments; 3, Differences between finance needed for plants in operation and for building new facilities.

684 Financing nuclear power projects: new investment powers of Euroatom, by O. Hahn. (In Nuclear Eng. Int., vol. 24 [285] May 1979, pp. 67-68)
A steady expansion of nuclear capacity in the European Community is one of the major ways of preventing unacceptable rises in oil imports. Over the period 1978 to 1985 the creation of the necessary nuclear capacity will require about £42,000 million, and to help utilities Euroatom is prepared to provide loans up to 20 percent of the total investment cost of the project. The purpose is to complement, not to replace, traditional financing resources.

685 Financing the Spanish nuclear power plant programme, by J. De Cubas. (In Nuclear Eng., vol. 20, Mar. 1975, pp. 182-183)
Tables: 1, Analysis of Spanish program; 2, Zorita 1 PWR station--loan negotiated by Exim Bank; 3, Typical phase 2 station Exim Bank loans.

686 Meeting the financial needs of the nuclear power industry, by D.M. Slavich and C.W. Snyder. (In Nuclear Eng., vol. 20, Mar. 1975, pp. 161-164)
The capital required to finance development of nuclear power stations and the associated fuel facilities during the

period 1971 and 1985 is estimated at $295 billion. This is not only significant in the comparison with historical levels of investment by electricity utilities, but in relation to the capital resources available. The high capital cost of nuclear stations adds to utilities' current financial problems but nuclear power offers long-term cost stability and financial institutions appear to be responsive to this opportunity.# Headings: Capital costs; Capital requirements; Financial sources.# Partial Charts: Estimated costs for nuclear power, 1968-1976; Estimated nuclear investment as a percentage of total investment in production plant; Energy production of US capital investments.# Tables: Electric utility industry capital requirements; Nuclear fuel cycle capital requirements; Total nuclear industry capital requirements.

687 Monogram helps estimate capital and operating cooling-system costs, by V. Ganapathy. (In Oil & Gas J. Tech., vol. 77 [17], Apr. 1979, p. 66)
Detailed estimates of the capital and operating costs for plant cooling systems can take time. A monogram can permit fast estimates. Chart shows cooling system costs.

688 Nuclear insurance shakes under guns of Denenberg. (In Elec. World, vol. 180, Sept. 15, 1973, pp. 27-28)
Abstract of a speech by Herbert J. Denenberg at nuclear power plant insurance hearing in Philadelphia.

689 Nuclear power for the production of synthetic fuels and feed stocks, by M. Steinberg. (In Energy Policy, vol. 5[1], Mar. 1977, pp. 12-24)
Partial Tables: Power plant generation costs, 1985 ...; Capital investment estimates: synthetic liquid fuel from coal and nuclear power; Production cost estimate; synthetic liquid fuel from coal and nuclear power; Production cost estimates synthetic liquid fuel from air, water, and nuclear power....

690 Ploy may like breeder funds. (In Elec. World, vol. 176, Aug. 15, 1971, p. 28)
A new ploy by the Joint Committee on Atomic Energy may provide the thrust necessary to complete the long awaited financial package for the first breeder reactor demonstration plant.

Management

691 Application of plant outage experience to improve plant performance, by D.G. Bridenbaugh and G.D. Burdsall. (In Combustion, vol. 46, July 1974, pp. 20-21)
Describes nuclear power as the modern miracle--a clean, economical, almost limitless source of power, to satisfy the

energy needs with minimum impact to the environment and minimum dependence on the limited fossil energy resources. Paper presented at the American Power Conference, Chicago, April 29, 30, and May 1, 1974.

692 Breeder advocates look beyond Demo 1. (In Elec. World, vol. 181, June 15, 1973, pp. 48-51)
Framers of national energy policy recommend commercial breeder industry and favor liquid metal (LMFBR) technology. Second round programs focus on economies of fuel, component scale-up, safety.# Tables: Forecast of estimated uranium prices translated from AEC forward costs and based on best information on ore availability and consumption; Capital cost premium that can be paid for LMFBR over LWR and still achieve lower power-reduction costs.

693 Computer programs for the in-core fuel management of power reactors, by M. Schneeberger and A. Szeless. (In Nuclear Eng. Int., vol. 24, Apr. 1979, pp. 44-47)
As a result of considerable utility demand, the IAEA has prepared a survey of all the computer programs available to optimize in-core fuel management. Factors which went into this project are detailed.

694 Development in nuclear power economics: abstract, by P. Sporn. (In Elec. World, vol. 173, Feb. 2, 1970, p. 21)
Table: Nuclear vs. coal-generated electric energy--assumptions: Costs are estimated as of July 1, 1969 for nuclear plant to be completed in 1976; Fossil plant to be completed in 1975; Capital cost for 1,100-MW nuclear plant is figured at $203.50 per week....

695 High burn up from natural uranium: the Candu experience, by R.A. James and M. Matthias. (In Nuclear Eng. Int., vol. 24, Apr. 1979, pp. 39-41)
Natural uranium fuel, on-power reloading, and low element failure rate, all combine to give Candu reactors low fueling costs.

696 How to work with a consultant in setting up an industrial plant energy-conservation programme interview. (In Power, vol. 116, Sept. 1972, pp. 54-55)
A question and answer situation to improve energy savings in industry.

697 In-service inspection of nuclear power plants, by F.C. Olds. (In Power Eng., vol. 79, Jan. 1975, pp. 28-35)
Direct cost of in-service inspection and associated operations over plant lifetime may be in the tens of millions of dollars; Inadequate early planning by the utility will raise the tab. The ultimate benefit from the full, eventual scope of in-service inspection is hard to assess now but, for one

thing, it may resolve the argument over the probability of sudden and catastrophic failures in nuclear plants.# Tables: In-service inspection of nuclear plant; Instructions to prospective firm-price bidders showing the extent to which owner expected the ISI program to apply to Rancho Seco nuclear power plant.# Fire instructions from two sections are shown in this example.

698 Is on-load refuelling really cheaper? by L.R. Howles. (In Nuclear Eng. Int., vol. 24, Apr. 1979, pp. 41-44)
Because routine inspection and maintenance can be done during refuelling shutdowns, the long-term achievement of LWRs may be comparable to that of on-load refuelling reactors. After four "fuel cycle periods" the load factor of both PWRs and GCRs approaches 75 percent.

699 Nuclear: is a fair press possible? (In Elec. World, vol. 182, Nov. 15, 1974, pp. 137-138)
An attack on the press for always alleging that nuclear power plants waste billions of dollars in idle capacity, but fail to compare it to any other energy source.

700 Nuclear fuel cycle: what's happening today? by R.G. Schwieger. (In Power, vol. 117, Sept. 1973, pp. 29-36)
Soft uranium market, higher SWU costs, new enrichment contracts, gas centrifuge developments, serious talk on plutonium recycle, stringent quality control in fuel fabrication, practical approaches to in-core fuel management.# Partial Charts: Estimates of future cost trends for each segment of the fuel cycle are based on 1973 dollars ...; Fuel cycle cost analysis; Plutonium recycle projections, 1972-2000.

701 Quick access to nuclear plant data, by J.H. Nail. (In Power Eng., vol. 77, Dec. 1973, pp. 44-47)
Describes the computer-based communications network called NORAN (Nuclear Operations Reliability Assistance Network) established by Nuclear Assurance Corporation.

702 Records management for nuclear power plants. (In Power Eng., vol. 83[5], May 1979, pp. 62-67)
Discusses a records management system developed to keep documents generated during design and construction.

703 Risk analysis of nuclear waste management systems, by T.H. Smith and W.K. Winegardner. (In Power Eng., vol. 79, Feb. 1975, pp. 35-37)
A two-year study to assess the safety of proposed waste management system undertaken at Battelle Pacific Northwest Laboratories.

704 United Kingdom survey. (In Nuclear Eng., vol. 14, Sept. 1969, pp. 713-748)

An extensive survey with very little said about investment planning and cost.

705 Use of computers in nuclear fuel management, by W.W. Brandfod. (In IEEE Trans Power Apparatus & Syst., vol. 90, Nov. 1971, pp. 2534-2539)
Describes the nuclear fuel cycle and some of the problems of fuel management. A description is given of various physics and economics computer codes used to assist in extracting the minimum benefits from this energy source.

Statistics

706 Appraisal of nuclear power plant reliability, by G.H. Applegren and others. (In Power Eng., vol. 79, May 1975, pp. 45-47)
Commonwealth Edison finds that nuclear plants in its system are more economical than coal-fired plants even at capacity factors well below 65 percent, that its nuclear plants have slightly better availability than its fossil plants....# Tables: 1, Gross capacity and availability factors for Commonwealth Edison's large base load units; 2, Costs of elasticity from Commonwealth Edison's large base load units using 1974 operating data; 3, Total generating cost vs. factor for large base load units.

707 Nuclear power growth, pains persists, by F.C. Olds. (In Power Eng., vol. 77, Nov. 1973, pp. 54-61)
Tables: 1, Comparison of AEC forecasts for nuclear power plants in commercial operation, 1962-1972; 2, Forecasts other than AEC's for nuclear power plants in commercial operation, 1967-1973.# Chart: Various growth curves for target of 150 Gwe nuclear capacity in commercial operation by the end of 1980.

708 Nuclear power plants: what it takes to get them built, by R.N. Budwani. (In Power Eng., vol. 79, June 1979, pp. 38-45)
A vital first step in support of corrective action, as detailed here--that of breaking down the material, manpower, cost, and schedule performance of nuclear plants, both built and planned, from the late 1960s to the mid-1980s. Gives statistical details.

709 Nuclear power to ease energy crisis, by F. Felix. (In Elec. World, vol. 179, June 15, 1973, pp. 80-82)
Stepping up the current program by 50 percent would cut oil imports through 2005 by 67-billion bbl, saving $500-billion in off-shore payments while easing the environmental impact.# Tables: Nuclear units slated to dominate electricity generation, 1970-2005; Projected growth, thousands of

Mw, 1970-2005; Accelerated nuclear program saves more fuel; Accelerated nuclear program could cut oil imports.

710 Nuclear survey: orders and cancellations, by W. Mitchell. (In Elec. World, vol. 182, Oct. 15, 1974, pp. 41-50)
Mixed bag of statistics shows commitments to new units running about as predicted, but mid-year inflation forces caused widespread cancellations and delays in construction programs.

711 Nuclear survey: 1972 looms as biggest year, by S.D. Strauss. (In Power, vol. 117, Jan. 1973, pp. 21-25)

712 Optimal timing of the US breeder, by R.G. Richels and J.L. Plummer. (In Energy Policy, vol. 5[2], June 1977, pp. 106-121)
Charts: 1, U_3O_8 cost vs. supply estimates; 2, Electric energy costs; 3, Non-electric energy costs; 7, Rising R&D costs ($ billion discounted to 1975 at 10 percent per year). Table: R&D cost projections, 1975-2020 (billion undiscounted 1975 dollars)

713 What really happened in 1974 to nuclear growth schedules and schedule slippage, by F.C. Olds. (In Power Eng., vol. 79, Apr. 1975, pp. 49-51)
A more accurate evaluation of the nuclear industry's growth problems is provided by a ten-year review of what has happened to each individual plant year by year.# Tables: 1, Nuclear power plants ordered from 1965 through 1974, showing originally scheduled plant capacities, originally scheduled dates for commercial operative, and subsequent changes by years; 2, Nuclear plant slippages, postponements, and cancellations summarized by year.

714 World nuclear programmes, by G. Greenhalgh. (In Nuclear Eng., vol. 20, Mar. 1975, pp. 164-169)
Tables: 1, Changes in moving annual load factors ...; 2, Cumulative load factor as a function of reactor type; 3, Nuclear power stations in the Foratom countries under construction and order; 4, Planned nuclear power stations in the Foratom countries; 5, New nuclear stations construction starts planned in Japan over three-year period, 1974 to 1976; 6, The electric power generating capacity and the nuclear capacity for 1970 and projections for 1975, 1980, 1985 and 1990 for the Foratom countries.# Charts: Comparison of nuclear plant cost estimates (total investment costs for MW (e) units).

715 Year of the cutback: industry report, 1974-1975. (In Combustion, vol. 46, June 1975, pp. 10-24)
Statistics: A sampling of scheduled nuclear plant con-

struction, 1975-1985; Manufacture of nuclear power reactors --50,000 Kwe and large; Manufacture of nuclear steam turbine-generators--50,000 Kw and large.

VII. OIL

Costs

716 Accelerated development planned for Beatrice, by S. J. McTiernan. (In Petroleum Eng. Int., vol. 51[1], Jan. 1979, pp. 66-70)
Small oil companies play an invaluable role, even in hostile, high-cost environments such as the North Sea.

717 The billion dollar market. (In Noroil, vol. 7, Aug. 1979, pp. 37-40, 57)
Tables: 1, Operating cost distribution; 2, Northern North Sea operating cost, 1979; 3, North Sea operating costs; 4, North Sea costs related to inspection, maintenance and repair.

718 California drilling cost seen big advantage. (In Oil & Gas J., vol. 77[18], Apr. 1979, p. 130)
Compares production costs with those in Texas.

719 Can new equipment cost less? by F.C. Jelen and M.S. Cole. (In Hydrocarbon Process, vol. 50, July 1971, pp. 97-100)
The optimum service life, or when to replace, can be calculated by studying the annual rate of technological improvement for various types of equipment.

720 Cost of Benzene reduction in gasoline to the petroleum refining industry: final report, by F.C. Turner, J.R. Felten and J.R. Kittrey. Cambridge, MA: Arthur D. Little Inc., 1978. 247p. (PB-282 743)
Discusses the cost to US petroleum industry of removing Benzene from the two largest contributors to the levels in the gasoline post-refinery reformates and FCC gasoline.

721 The cost of North Sea oil development, past trends and future prospects, by N. Trimble. Aberdeen: University, Department of Political Economy, 1976. [3], 14p. (North Sea Study. Occasional Paper, 10)
Partial Contents: 2, Past trends in development costs; Future prospects for development costs.

722 Crude oil dehydration: a look at methods and costs, by H.

Wallace. (In World Oil, vol. 189[6], Nov. 1979, pp. 73-74, 76)
Discusses the basic operating and cost factors to be considered when selecting an energy-efficient treating system for oil/water emulsions.

723 Developing new oil may cost $70 billion annually. (In World Oil, vol. 189[6], Nov. 1979, pp. 129-130, 132, 137)
The total annual capital investment required by the world petroleum industry to develop new oil could move up from over $70 billion next year to more than $110 billion per year in the 20 years thereafter. The cause for such increases is clearly the much greater cost of developing the most expensive category of new oil from Arctic ventures....

724 Drill string stabilization cuts shallow drilling costs, by H.J. Smith and R.W. Scott. (In World Oil, vol. 189[5], Oct. 1979, pp. 55-58)
Partial Tables: Effect of bit type on ROP and cost per foot; 3, Effect of bottom hole assembly on ROP and cost per foot; 4, Effect of stabilization on ROP and cost per foot.

725 Enhanced oil recovery: tools and techniques of costly quest, by A.W. Crull. (In Energy, vol. 5[1], Winter 1980, pp. 12-13)
An estimated 300 billion barrels of petroleum in US reservoirs are unrecoverable by conventional methods. A respectable amount can be recovered if money is to be spent on it.

726 Fallacy of lighting energy conservation: a look at consumer usable fuels, by K. Southward. (In Lighting Design & Appl., vol. 5, Mar. 1975, pp. 31-33)
Reducing lighting consumption across the board by 50 percent would cost an additional 0.30 percent of the nation's total energy consumption in consumer usable oil and gas.

727 Find optimum furnace efficiency, by S. Menicatti and L. Cappiello. (In Hydrocarbon Process, vol. 51, Sept. 1972, pp. 226-230)
Describes a technique which permits furnace optimization on an annual basis by including depreciation, maintenance and operating costs.# Partial Charts: Cost of radiation coil and costing versus process duty; Costs of transportation and installation versus process duty; Fuel costs at various steam economies versus generation (a) and air preheating (b) at five years' depreciation; Fuel costs at various steam economies versus generation (a) and air preheating (b) at two years' depreciation; Optimum efficiency versus fuel costs for two and five years' depreciation; Total annual optimum cost versus fuel cost for two and five years' depreciation.

728 Forecasting offshore platform development costs, by J. D. Culver. (In Pet Eng., vol. 44, May 1972, pp. 79-82, 84)
Approximately $10 billion have been spent in the Gulf of Mexico alone searching for oil and gas. In order to continue safe operations, and to do so at a profit, factors that eventually will shape the financial outcome of a project must be forecast as accurately as possible.

729 Further rise in oil costs. (In Pet. Econ., vol. XLVI[7], July 1979, p. 268)
Report on the OPEC meeting at Geneva on 26th June 1979.

730 Guidelines in cost control--1: people cause control production costs, by C. F. Dwyer. (In Pet. Eng., vol. 44, Aug. 1972, pp. 52, 56-57, 60, 62)
People, not operational difficulties, are the real problem in cost control. Successful cost control occurs only where there's a flexible, positive, "can do" attitude on the part of management.

731 How one independent cuts Gulf Coast drilling costs, by G. Brown. (In Oil & Gas J., vol. 77[44], Oct. 29, 1979, pp. 134, 139-140)
Includes cost comparisons and a table showing cost analysis.

732 Local cost control improves rig profits, by R. C. Parsons. (In World Oil, vol. 188[7], June 1979, pp. 181+)

733 Low-colloid oil muds cut drilling costs, by J. P. Simpson. (In World Oil, vol. 188[5], Apr. 1979, pp. 167-170)

734 New approach to preliminary cost estimating, by J. B. Swaney. (In Hydrocarbon Process, vol. 52, Apr. 1973, pp. 167-169)
Contends that if you know the delivered cost of the process equipment you can use the curves and tables presented in this paper to find direct labor costs, equipment and tool costs.# Curves: Process equipment costs vs. direct labor manhours for four processing units; Direct labor manhours vs. construction time; Direct labor manhours vs. direct labor manhours; Percent of construction time vs. percent of average manpower.# Tables: 1, US labor productivity multipliers; 2, Process unit craft manpower distribution at average work force; 3, Direct labor costs; 4, Labor, equipment and tool cost summary.

735 New controller optimizes pumping-well efficiency, by J. K. Godhey. (In Oil & Gas J., vol. 77[39], Sept. 24, 1977, pp. 161-165)
Using its sensitive automatic liquid level monitor, Mobil Research has designed an improved new generation of the pump-off controller. It reduces operating cost and opti-

mizes production in rod-pumping or submersible-pumped wells.

736 North Sea costs escalation study: vol. 2, final report. London: Department of Energy, 1975. 1v. (variously paged)
A study undertaken by Peat Marwick, Mitchell & Co., management consultants.# Contents: Pt. A, The nature and extent of cost escalation--oil and gas exploitation in the North Sea, the measurement of cost escalation, analysis of cost escalation experience, comparison with other major developments, the causes of escalation; pt. B, Subsystems--resources, project management, cost escalation (the oil companies perspective); pt. C, The future.

737 North Sea costs escalation study, vol. 1, by the Department of Energy. London: HMSO, 1976. vi, 126p.
Partial Contents: The definition of cost escalation--(Analysis of cost escalation, Accuracy of cost escalation, Hidden escalation ...); Cost escalation experience (cost indices); Causes of escalation--input cost movement; The effects of cost escalation (The effect of future cost movements on Government revenues, Effect of cost escalation on licensee, Revenues and finance); Future cost escalation.

738 North Sea oil: a review, by J. Fernie. Huddersfield: The Polytechnic Department of Geography & Geology, 1976. [24]p. (Occasional Paper 4)
Partial Contents: The escalation of costs; Sources of finance.

739 North Sea oil and gas: a geographical perspective, by K. Chapman. Newton Abbot: David & Charles, 1976. 240p.
Partial Contents: Pt. 1, Offshore patterns--the pattern of development: variables influencing costs, variables influencing revenue.

740 North Sea oil: the great gamble, by B. Cooper and T.F. Gaskell. London: Heinemann, 1966. [8], 179p. illus.
Includes cost considerations of exploration and production, drilling operations, helicopters, platforms and transport.

741 Oil drilling programs stimulant to oil development, by F.M. Stewart. (In J. Pet. Tech., vol. 23, May 1971, pp. 539-545)
Contends that a common characteristic of drilling programs is that prospects to be explored or drilled are not specifically defined at the time the program is formed. The sponsor (operator or management company) furnishes the organization and know-how, investors supply most of the capital. Cost and revenues are allocated between management and investors according to a specified sharing arrangement.

742 Oil shale technologies are ready for commercialization, by S.H. Zukor and P.A. Petzrick. (In Energy, vol. 5[3], Summer 1980, pp. 13-15, 20)
Oil shale is being recognized more widely as the best and lowest cost alternative to petroleum derived fuels.

743 Operating costs and performance of Large-volume submergible pumps, by R.K. O'Neil. (In J. Pet. Tech., vol. 24, Dec. 1972, pp. 1473-1478)
Presents some statistics on the operating costs and performance of a large number of pumps.# Partial topics: Repair and maintenance costs; Field experience--repair and exchange costs; Calculation of theoretical operating costs.

744 Optimizing for lower cost-per-foot drilling, by R.M. Robson. (In World Oil, vol. 189[4], Sept. 1979, pp. 81-83)
Offset well data and a computer analysis are used to plan the best possible procedure for drilling a proposed well.

745 Pipeline risers plan now for cost savings, by R.J. Brown. (In Noroil, vol. 7, Apr. 1979, pp. 65, 92-94)
Discusses a proven new technique for the tricky task of the subsea connection of pipelines to platform risers. Forward planning is the key to the significant cost reductions which it offers.

746 Placid sees lower costs with Subsea completion, by R.B. Buron. (In Pet. Eng., vol. 41, Aug. 1969, pp. 56, 58 and 62)
Discusses the long-range maintenance cost of landmark mile stepout well in the proven Ship Shoal area of Louisiana.

747 Proper use of diamond bits can reduce drilling costs, by J.H. Striegler. (In World Oil, vol. 189[1], July 1979, pp. 109-110, 112, 114, 116)
Discusses guidelines used for applying diamond bits and details operating experience that led to the significant reduction in costs.

748 Subsea production systems outlook. (In Oil & Gas J., vol. 77[38], Sept. 17, 1979, pp. 97-98, 100, 102)
Presents a table showing cost comparison between a subsea and a platform well.

749 Superintendent eyes ways to cut costs. (In Pet. Eng., vol. 44, Aug. 1972, pp. 64, 66, 68)
Independent producers in the past decade have been under pressure to continually find new ways to cut production costs through better planning and more efficient engineering.

750 $330-million for downhole repairs, by J.E. Kastrop. (In Pet. Eng., vol. 44, July 1972, pp. 37-40)
Discusses the cost of downhole equipment repairs and

replacements in half a million US oil producing wells.# Partial Headings: Basis for downhole costs; Cost records aid profits.# Charts: Relationship by lift method of yearly downhole costs to maintain US oil wells; Annual material expenditures for downhole repair/replacement in US oil wells last year.# Tables: Downhole costs to maintain US oil producing wells in 1971.

751 Trends in the Soviet oil and gas industry, by R.W. Campbell. Baltimore, London: Johns Hopkins, 1976. xvi, 125p. illus.
Partial Contents: Review of Soviet energy policy--the economic rationale; Oil production--cost of production; Economic reform in the oil and gas industry--pricing and costing; Fuel allocation models and shadow prices.

752 US crude costs soar. (In Pet. Econ., vol. XLVI[9], Sept. 1979, pp. 376-377)
Table: USA--origins of direct crude imports and indicated average weighted prices.

753 Utilize utility dollars, by M. Brooks. (In Combustion, vol. 42, Sept. 1970, pp. 22-24)
Discusses money and its management in an oil refinery.# Table 2, Money value and depreciation factor.

754 What are offsite costs of processes? by W.L. Nelson. (In Oil & Gas J. Tech., vol. 77[20], May 14, 1979, pp. 146, 150)
An editorial answer to the question: "In some of your articles on process unit costs ('Crude distillation,' Feb. 25, 1974, etc.) you suggested 'process of site' costs that are excessively large. What is the explanation?".# Table: Offsite investments (1973) for process operations.

755 What will future refineries cost? by E.K. Grigsby and others. (In Hydrocarbon Process, vol. 52, May 1973, pp. 133-135)
Using a hypothetical refinery and assumptions about growth crude sources transportation and distribution total US refining capital required is $2,000/bpd.# Tables: 1, Refinery charge and yield; 2, Capital requirements; 3, Range of operating costs; 4, Crude oil cost at refinery; 5, Cost of refined products at the refinery; 6, Cost of distribution of refined products. Includes a product cost equivalent chart.

756 Where do funds fit in future drilling? by G.C. Hann. (In Pet. Eng., vol. 43, Sept. 1971, pp. 66, 68, 74)
"It is probable that as many as one in three domestic wells drilled during the next ten years will involve funds assembled by a mature and responsible investment program industry--funds which otherwise would probably not be used in oil and gas exploitation. This could be a significant factor in the planning of the companies who will be called upon to do the drilling."

757 Who follows OPEC? by J. Maddox. (In Chem. & Ind., June 1, 1974, pp. 424-425)
Discusses OPEC and its monopoly of cheap oil.

758 Workover economics--complete but simple, by J. L. Rike. (In J. Pet Tech., vol. 24, Jan. 1972, pp. 67-72)
Presents a simple set of tables devised to give profit-to-investment ratio and rate of return, using the payout figure usually calculated, but further taking into account acceleration economics, risk, and anticipated life. The tables can be applied to any venture having a single initial cost and a consistent (not necessarily constant) type of income return.

Economics

759 Analysis of Appalachian Basin economics, by J. G. Redic. (In J. Pet Tech., vol. 26, July 1974, pp. 717-723)
Contends that with improved wellhead prices, substantial risk capital is now available for developing modest per-well reserves in the Appalachian Basin.# Topics: Considerations in determining Appalachian Basin investments and profitability; Determination of average costs, rates of production, reserves; Leasing of tangible well equipment; Estimate of costs and revenues; Undiscounted profit and rate of return.# Tables: 1, Determination of net cash investment; 2, Example: production forecast and cash flow projection ... nonlease case; 3, Example: production forecast and cash flow projections ... (constant payments).

760 Dry and wet subsea system economics, by C. S. Kubena. (In Pet. Eng. Int., vol. 51[5], Apr. 1979, pp. 66, 68, 72, 74)
Headings: Wet system; Dry system; Wet tree maintenance; Insurance and risk factors; Installation times; Price and installation comparison; Dry system design and economics.# Tables: 1, Dry system maintenance costs at various water depths; 2, Wet tree maintenance costs at various water depths.# Charts: Percent difference in cost of dry system vs. wet system; Total averaged hardware, installation and maintenance costs of dry system vs. wet system.

761 Gulf of Alaska development hinges on giant reserves, by W. W. Wade and P. Hanley. (In Oil & Gas J., vol. 77[38], Sept. 17, 1979, pp. 111-115)
Two-thirds of the US outer continental shelf adjoins Alaska, making this a future primary leasing area. The economics of developing any reserves, especially in the Gulf of Alaska, are essential to measuring the worth of such leases.# Partial tables: Minimum field sizes--(5 price required to earn 15 percent for 20-25 yr. field [$/bbl] million bbl); Investment.

762 Independents facing severe cash flow drains. (In Oil & Gas J., vol. 77[15], Apr. 9, 1979, p. 25)
Chart shows how percentage depletion rate is dropping.

763 Inflation, dollar depreciation, and OPEC's purchasing power, by M. Dailami. (In J. Energy Dev., vol. 4[2], Spring 1979, pp. 336+)

764 Motor gasoline supply and demand, 1967-1978, by K. E. Seiferlein. Washington, D. C.: Department of Energy, Energy Information Administration, 1978. 36p.

765 National oil account, 1976/77. London: HMSO, 1978.
Issued annually in pursuant to section 40(4) of the Petroleum and Submarine Pipeline Act 1975.

766 North Sea oil: its effect on Britain's economic future, by I. Everingham and others. Fontainebleau: European Institute of Business Administration, 1975. [2], 49 leaves.
Partial Contents: The contribution of North Sea oil to the UK energy balance; The effect of North Sea oil on industry and Scotland; The financial aspects of North Sea oil --capital requirements, techniques and sources of finance, impact of North Sea oil on Britain's balance of payments and exchequer revenue.

767 Oil politics in the 1980s: patterns of international cooperation, by C. Noreng. New York: McGraw-Hill, 1978. 171p.
Published as part of 1980s Project of the Council on Foreign Relations.

768 Oil revenues and accelerated growth: absorptive capacity in Iraq, by K. A. Al-Eyd. New York: Praeger, 1979. xiv, 188p.

769 OPEC's share in downstream operations: the transportation case, by F. Azarnia and I. Bejarano. (In OPEC Rev., vol. 3[2], Summer 1979, pp. 74+)

770 Power from the sea: the search for North Sea oil and gas, by C. Callow. London: Gollancz, 1974. 190p. illus.
Partial Charts: Oil is the prize; Now the money men move in.

771 US overseas wages: are they comparative? by R. Gant. (In World Oil, vol. 189[4], Sept. 1979, pp. 87, 89, 91-92)
The US expatriate's financial situation was improved by the 1978 tax law, but more progress is still needed. # Partial Tables: Tax liability computation; Adjustable gross income; Adjustable gross income less deduction; Taxable income.

772 World oil and OPEC: the razor's edge, by M.F. Thiel. (In World Oil, vol. 189[5], Oct. 1979, pp. 123-124, 126, 128, 130, 133)
An assessment of the economic and political situation reveals important policy considerations.

Finance

773 All set for development. (In Noroil, vol. 7, Aug. 1979, pp. 19-21)
Several UK and Norwegian fields are now set for development, with orders totalling more than six billion dollars.

774 Annual report survey, 25th, by E. Adams. (In Pipeline & Gas J., vol. 198, May 1971, pp. EM9-16)
Chart: Earnings of petroleum companies. # Tables: 1, Financial statistics of US petroleum companies; 2, Financial statistics of natural gas companies; 5, Oil and liquids producers among US companies; 6, Refining operations by companies; 7, Petrochemical sales; 8, Natural gas pipelines; 9, Refined petroleum products sales; 10, Free world expenditures for facilities and reserves.

775 British banks compete for share of oil business. (In Offshore Eng., Oct. 1979, pp. 63+)

776 Capital needs in the 1980's. (In Pet. Econ., vol. XLVI[7] July 1979, p. 276)
Because oil is capital-intensive, high-risk industry, the continuing expansion necessitated by the growing needs of oil consumers will absorb a great deal of capital.

777 Capital outlays due for rise in 1971. (In Pet. Eng., vol. 42, Aug. 1970, pp. EM7-9)
Tables: Free World expenditures for facilities and reserves for 1969-70-71; Total oil and natural gas outlays--natural gas outlays, oil outlays; Earnings declined 20 percent, 1951-71.

778 Capital short? follow one of these directions, by L.A. Murphy. (In Pet. Eng., vol. 42, Nov. 1970, pp. EM5-8)
When an oil and/or gas company needs new funds for improvements or expansion, the most beneficial source of such funds is its accumulated net income. Net income represents the profit that the company has earned on which it pays no interest and with no dilution of stockholders equity and no reduction of the company's assets. Discusses corporate bonds, venture capital, unsecured loans, leasing and sale of property.

779 Conference on North Sea, 1&2, 1972. London: Financial Times, 1972. 137p. illus.

Presents papers read at the conferences sponsored by the Financial Times, Investors Chronicle and Petroleum Times. # Partial Contents: The financing of the North Sea exploration--an American view, by W. Spencer; Some investment implications of North Sea oil and gas--a Scottish view, by I. Noble; The role of the City in financing North Sea development, by F.J. Hervey-Bathurst.

780 [Conference] on oil in deeper waters: speakers' papers. Kingston upon Thames, Surrey: Spearhead Pubs., 1976. 148p.
Partial Contents: The World Bank looks at oil in deeper waters, by L.C. Peacock; Financing North Sea and European offshore dependents, by J.A. Adams.

781 Cornwall and the offshore oil industry: a directory for business enterprise. Camborne, Cornwall: Cornish Institute of Engineers, 1974. 1v. (variously paged)
Includes an analysis of financial involvement.

782 Current developments in petroleum financing, by T.G. Stevens. (In J. Pet. Tech., vol. 23, Feb. 1971, pp. 202-204)
Contends that the event that has caused the greatest change in petroleum financing is the tax reform act of 1969. Principally, the major losses to the petroleum industry because of this law are the discontinuance of production payment financing and the reduction in percentage depletion.

783 Effect of UK oil reserves. (In Pet. Econ., vol. XLVI[7], July 1979, p. 288)
Table: Economic profile of the UK North Sea oil, 1973-85 (figures in £billion, current)

784 Enhanced recovery financial aid seen needed. (In Oil & Gas J., vol. 77[18], Apr. 30, 1979, p. 128)
Gary Operating Co., Englewood, Colo. contends that government financial incentives beyond global market prices for crude oil are needed to stimulate investment in US enhanced oil recovery projects.

785 Evaluation by rate of return of present value? by C.E. Haskett. (In Pet. Eng., vol. 44, Aug. 1972, pp. 48-50)
Headings: Discounted cash flow; Average annual rate; Modified rate of return; Present value criteria. # Charts: Discounted cash flow rate of return; Average annual rate of return; Modified rate of return; Discounted (10 percent) net operating revenue to investment ratio; Percentage gain on investment.

786 Evaluation problems as related to Appalachia bank financing, by J.G. Redic. (In J. Pet Tech., vol. 22, Oct. 1970, pp. 1291-1298)
In the Appalachian area, common engineering techniques are not applicable in evaluating oil and gas reserves for

bank financing, principally because of a simple lack of necessary reserves data. The empirical methods discussed have shown themselves to be reliable when used with judgment and discretion.

787 Facing up to an energy crisis: an OPEC view, by R.G. Ortiz. (In OPEC Bull., Sept. 1980, pp. 7-13)
Contends that OPEC would like to see a floor-price containing the following two elements which are considered as the major factors affecting the purchasing power of the oil revenues: 1, An index reflecting the impact of inflation on international trade, both merchandise and services; 2, An automatic exchange rate adjustment factor based on the main currencies used in world trade.

788 Finance, by P.H. Richards. (In Pet. Rev., vol. 33[395], Nov. 1979, p. 15)
Shell is taking over a Californian oil company which owns huge unexploited reserves, but the price will be very high.

789 Finance and investment: fighting back the money flood. (In Pet. Econ., vol. XLVI[11], Nov. 1979, pp. 453-454)
Headings: Outburst of speculation; Handsome profits; Market fluctuations. # Chart: Price indices for oil, gold and silver, 1973-1979 (1975=100).

790 Finance and investment: investment clouds gather. (In Pet. Econ., vol. 46[6], June 1979, p. 231)
The common assumption that petroleum ventures are among the most attractive of all investment opportunities is apparently being revised.

791 Finance: does UK policy provide the best climate for North Sea development or hinder progress by imposing PRT and fixed pricing? by P.H. Richards. (In Pet. Rev., vol. 33 [392], Aug. 1979, p. 21)
Discusses the effects of oil price increases on profitability of North Sea oil discoveries.

792 Finance: first quarter for BP. (In Int. Pet. Times, vol. 83, June 15, 1979, p. 8)
Tables: BP Group income statement; Capital expenditure of the BP Group.

793 Financing oil development. (In Noroil, vol. 7, Dec. 1979, pp. 59, 61, 64-65, 67)
Despite the daunting amounts involved the money will be available. The question is from where will the money come and at what price. # Headings: The buyer's market; Profit and investment; Outside finance; Project financing; The price of money; Government money. # Graphs: E & P capital expenditure; Shell funds and how they were used; Relative investment costs of various energy sources.

794 Foreign operations boost oil earnings. (In Oil & Gas J., vol. 77[33], Aug. 13, 1979, pp. 52-53)
Table shows how 25 companies fared financially the first half of 1979.

795 Global industry faces soaring capital needs. (In World Pet. Cong., no. 2; Oil & Gas J., vol. 77[39], Sept. 17, 1979, pp. 55-57)
Contends that the cost of developing new global oil production could rise from $2,000/b/d, the average of today's "low cost" oil, to $33,000 by 2,000.

796 Industry is spending more, but controlling less crude. (In World Oil, vol. 188[5], Apr. 1979, pp. 187-188)
Charts: Source of revenue, 1977; Costs, direct taxes and net income as percent of gross operating revenue; Revenue increase in 1977; Changes in 1977. # Tables: 1, Source of crude oil; 2, Capital expenditures.

797 North Sea revenue options. (In Pet. Econ., vol. XLVI[7], July 1979, pp. 266-268)
Discusses the substantial benefits from vast commercial investments in the North Sea. # Headings: Government revenues; Exporting revenues.

798 The north Sea: the search for oil and gas and the implications for investment. London, New York: Cazenove & Co., 1973. 1v. (variously paged) illus.
Partial Contents: Pt. 1, Exploration and development costs; pt. 3, Implications for European energy--European energy market, Regional economic repercussions, Investment implications. # Appendices include energy statistics--1937, 1950-1969.

799 Offshore development. Vol. 3: finance, taxation and government. London: Financial Times, 1975. [8], 163p.
Partial Contents: The United States energy situation--balance of payments and other financial and economic problems, by C. King Mallory; Will inflation produce insuperable difficulties, by J.G.S. Longcroft; Does the threat of government participation in UK North Sea oil hinder production financing? by E.E. Monteith; Energy finance in South East Asia--a merchant banker's view, by J.P. Silcock; Rising cost of offshore exploitation and production operations, by Bart Collins; Finance and offshore development, by B.E. Ramfors; Is financing North Sea support service companies still a major business prospect? by V. Lall.

800 Offshore North Sea technology conference and exhibition. 1974. Stavanger: Economy, Law, Energy, Marketing, [1974]. 1v. (variously paged)
Partial Contents: Capital requirements of energy supply, by E. Symonds; Exploration and development of the North

West European continental shelf: methods of finance and sources of funds, by M.J.K. Belmont; Economics of the North Sea ventures, by A. Hols.

801 Offshore Spain investment increases, by R. Richards. (In Ocean Industry, vol. 14[2], Feb. 1979, pp. 106+)

802 OPEC aid, the OPEC fund, and cooperation with commercial development finance sources, by I.F.I. Shihata. (In J. Energy Dev., vol. 4[2], Spring 1979, pp. 291+)

803 OPEC's soaring revenues. (In Pet. Econ., vol. XLVII[6], June 1980, p. 243)
Table: OPEC oil exports and revenues, 1974-1979.

804 Post-1973 oil investment patterns in producing countries, by F. Al Chalabi. (In OPEC Rev., vol. 3[1], Mar. 1979, pp. 8+)

805 Private capital aids [Argentina's] oil drive. (In Pet. Econ., vol. 46[10], Oct. 1979, pp. 409-412)
Oil production has risen sharply owing largely to the reopening of exploration and development to private capital. # Table: Argentina--Petroleum statistics.

806 Probability models for petroleum investment decisions, by M.B. Smith. (In J. Pet. Tech., vol. 22, May 1970, pp. 543-550)
The time, effort and money required to develop a computer program for a general probabilistic economic model are great, but a single investment decision made more sound by the use of the model can more than justify the expense.

807 Production spending doubles, other spending stable. (In World Oil, vol. 189[4], Sept. 1979, pp. 108-109)
Between 1967 and 1977, US production expenditures rose 11 percent while outlays for downstream projects grew 11 percent. # Charts: Worldwide capital and exploration expenditures, 1967-1977; Worldwide capital expenditures, 1977.

808 Profitability analysis of leveraged transactions, by M. Silberg and F. Brons. (In J. Pet. Tech., vol. 25, Mar. 1973, pp. 319-328)
Because the decision function is usually separated from the financing function, most venture analyses do not encompass the source, the cost, or the repayment of the necessary capital. This must change as the petroleum industry moves toward more venture financing, for in leverage--or "project financed"--ventures, those factors cannot be ignored. Tables: 1, Discounted profit on equity--three kinds of $1,000 investments; 2, Sample projects; 3, Cash flow patterns--equity funds; 4, Relationship of revenue and profit as performance departs from best estimate for ventures

requiring $1,000,000 equity plus varying loans; 5, Analysis of uncertainty in performance of case 1, table 4; 6, Analysis of uncertainty in performance for all four cases of table 4; 7A, Discounted value of five equal annual (lease) payments; 7B, Segregation of annual payments on A $1,000 loan into interest and principal; 7C, Lease payments discounted at earning-opportunity rate; 7D, After-tax leasing case.

809 Qualifying HPI uncertainties: special report, by J.N. Fisher and A.J. Phipps. (In Hydrocarbon Process, vol. 50, Mar. 1971, pp. 63-70)
Risk analysis gives new prospectives for reviewing opportunities and problems in hydrocarbon processing industries (HPI) for the 1970's. # Tables: Probable effects of pollution controls and materials price changes on product values by 1974/75; Probability distribution used for raw material costs on both a cumulative percentage and probability density function plot for 1975 raw material costs versus 1969-1970; Frequency plot on motor gasoline manufacturing costs; Frequency plot on benzene expected average cost, cents per pound, 1974-75; Frequency plot of ethylene expected average cost, cents per pound, 1974-75.

810 UK offshore prospects. (In Noroil, vol. 7[5], May 1979, pp. 91-93, 95, 97)
Chief constraint upon investment in oil and gasfields development in the United Kingdom sector of the North Sea remains lack of resources, mainly in terms of skilled manpower. # Tables: 1, Total expenditure in the UK and Norwegian North Sea sector, 1979-1987; 2, North Sea operation costs, 1979-1985.

811 US oil companies: earnings soar to record levels. (In Pet. Econ., vol. XLVII[3], Mar. 1980, pp. 123-124)
Tables: US companies--net income, million dollars; Earnings analysis--major groups.

812 Where the money will go! Total expenditure on offshore oil and gas will increase enormously in areas throughout the world. (In Noroil, vol. 7, Dec. 1979, pp. 33, 35, 37)
An offshore market forecast. In 1980, investment in developing new oil and gas production is calculated to exceed US $25 billion.

813 Whitehall slammed over North Sea oil licensing, by D. Booth. (In Engineer, vol. 236, Mar. 1973, p. 11)
Contends that Britain has sold its oil and gas too cheaply compared with other oil producing countries--a conclusion of the report of the Committee of Public Accounts on North Sea Oil and Gas.

Management

814 After the second shock: pragmatic energy strategies by R. Stobaugh and Daniel Yergin. (In Combustion, vol. 50, June 1979, pp. 8-21)
Examines, among other topics, how politically precarious the supply of foreign oil is and the economic and other costs of growing dependence.

815 Bridging the HPI interface, by R. L. Mitchell. (In Hydrocarbon Process, vol. 49, July 1970, pp. 83-85)
Management philosophy separates refining and petrochemicals. Bridge this interface for better use of capital and raw materials and increase future profits.

816 Divers--war under water. (In Noroil, vol. 7, July 1979, pp. 27, 29-30)
A slump in exploration and construction in the North Sea triggered fierce competition, aggressive price cutting and coincided with fundamental changes in the structure of the diving industry.

817 The effective management of resources: the international politics of the North Sea, edited by C. M. Mason. London: Francis Pinter, 1979. [10], 268p. illus.
Partial Contents: Resource management and international politics; Oil and gas: the international regime; Oil and gas: national regime; Oil and gas: governments and companies.

818 Exploring the Americas. (In Noroil, vol. 7, Dec. 1979, pp. 49, 51, 53, 55-56)
Throughout the hemisphere a great deal of money is being spent to develop existing reserves and find new ones.

819 First quarter profits show sharp jump. (In Oil & Gas J., vol. 77[21], May 21, 1979, pp. 28-29)
Presents figures for first quarter 1979 profits for the US oil industry.

820 Hondo drilling delay becoming more expensive. (In Oil & Gas J. Tech., vol. 77[30], July 23, 1979, pp. 48-50)
Investment continues to increase, with no return, due to regulatory red tape.

821 How Husky makes a profit in heavy oil production, by D. O. Gurel. (In World Oil, vol. 189[4], Sept. 1979, pp. 63-67)
Four different pumping methods are used to obtain a reasonable rate of return.

822 The management of North Sea projects, by Y. Arnoni. (In Noroil, vol. 7[2], Feb. 1979, pp. 79, 81, 83, 85, 87)
Compares management of large offshore projects and landbase projects. Cost is said to be multiples more ex-

pensive for offshore projects due to: Structure being 100s of feet high; Space limitations and congested work areas; Premium payments for offshore labor; High cost of plant mobilization; Weather down-time; Inefficiencies of access; Need for sufficient life support system; Environmental restrictions.

823 The North Sea market. (In Noroil, vol. 7[5], May 1979, pp. 81-88)
The North Sea oil and gas business has reached a level of around US $6.5 billion annually. A market worth approximately US $6 billion annually is sustainable over the long term. # Tables: 1, North Sea field development expenditures, 1979-1989; 2, North Sea exploration and appraisal drilling expenditures, 1979-1989; 3, North Sea field operating costs, 1979-1989.

824 Planning of refining/petrochemical complexes, by H. L. Kephart. (In Chem. Eng. Prog., vol. 67, Nov. 1971, pp. 68-74)
Reviews a technique for overcoming the reluctance of company personnel to use in-house expertise: that is, to transfer engineers from one department to another more frequently. # Charts: Organization of the Sun Oil Co.; Long-range planning cycle; Short-range planning system; Frequency of use, and quality evaluation sources by engineers. # Table: Overall ranking of information channels in descending order of importance.

825 Platform rig survey. (In Offshore, vol. 39[8], July 1979, pp. 65-66, 70, 72-74)
Operators, drilling sites and operating capacities are a few of the up-to-date statistics offered in this international survey.

826 Profit planning in the HPI, by W. L. Kephart and W. D. Trammell. (In Hydrocarbon Process, vol. 49, July 1970, pp. 91-98)
Management science can cope with the major uncertainties as today's highly interactive hydrocarbon processing industry's (HPI) plans for future profits.

827 Project performance evaluation, by R. F. Hartman. (In Chem. Eng. Prog., vol. 67, Dec. 1971, pp. 42-46)
The performance ratings of projects undertaken indicates that the evaluation system described is highly effective in stimulating increased output from the overall work effort.

828 A proposal for a new US oil policy, by A. E. Safer. (In Energy Policy, vol. 6[1], Mar. 1978, pp. 2-13)
Proposes the substitution of more involvement by the US government in international oil trade for less government in regulation of the domestic oil business. # Partial Headings: Rationalizing the crude oil market; A new fund for develop-

ing countries; Benefits and costs. Table shows payments for US domestic and imported oil, 1976-80.

829 Refining industry of the future, 1985. (In Pet. Chem. Eng., vol. 42, Mar. 1970, pp. 16-20, 22, 24-25, 28, 30-31)
Engineers and industry experts take a crystal ball look at a year which most trend factors point to as a pivotal one for the future of petroleum and other energy sources.

830 Study of potential benefits to British industry from offshore oil and gas development, by International Management and Engineering Group of Britain Limited. Department of Trade and Industry. London: HMSO, 1972. 136p.
Includes a chapter on financial and legal aspects; and a financial recommendation which favors cheap credits available to British industry.

831 World oil market data. Houston, Tx. World Oil. 159+
Annual. Contains data on production, trade and prices in the world oil market.

832 The world oil market in the years ahead: a research paper. Washington, D.C.: National Foreign Assessment Center, 1979. 80p. (ER79-10327U)

Pricing

833 An economic analysis of crude oil price behavior in the 1970's by W.J. Mead. (In J. Energy Dev., vol. 4[2], Spring 1979, pp. 212+)

834 Canadian self-sufficiency price tag seen high (In Oil & Gas J., vol. 77[45], Nov. 5, 1979, pp. 40-41)
The Canadian Petroleum Association contends that exploration and production spending of $200 billion and oil price increase of at least $4/bbl/year will be needed if Canada is to achieve self-sufficiency in oil by 1990.

835 A combined decline-curve and price analysis of US crude oil production, 1968-76, by A.E. Bopp. (In Energy Econ., vol. 2[2], Apr. 1980, pp. 111-114)
An approach combining decline-curve and price analysis is used to mode US oil production over the period 1968-76. # Tables: 1, Real wholesale price index for US crude oil and US domestic crude oil production levels; 2, US oil production levels.

836 Crude oil import prices, 1973/1978, by the International Energy Agency. Paris: OECD, 1979. 18p.
Gives statistical information.

837 Crude oil prices. (In Pet. Times, vol. 83[2109], Oct. 15, 1979, p. 12)

A tabulation of state selling prices for the world's major export crudes, 1977-1979.

838 Current world crude price trends. (In Pet. Times, vol. 83, June 1, 1979, p. 30)
Gives tabulated statistical information on prices.

839 Economic impacts of a transition to higher oil prices, by R. G. Tessmer, and others. Upton, N.Y.: Brookhaven National Laboratory, 1978. 64p.

840 EEC demand for imported crude oil, 1956-1985, by G. Kouris and C. Robinson. (In Energy Policy, vol. 5[2], June 1977, pp. 130-141)
Partial Tables: 3, Oil consumption, relative prices and real GDP in EEC; 4, Per capita income, price and temperature elasticity--overlapping decades; 5, Price changes required to hold EEC net oil imports constant (British North Sea crude supplies 150 million tonnes in 1985)

841 ERA revamps gasoline price, allocation regs. (In Oil & Gas J., vol. 77[30], July 1979, pp. 22-23)
Comment on markup allowed to US gasoline dealers over their acquisition cost under new regulation issued by the Economic Regulatory Administration.

842 European pricing policies. (In Pet. Econ., vol. XLVII[4], Apr. 1980, pp. 142-143)
Headings: Controlled markets; Effects of control; Average realizations.

843 Factors affecting world petroleum prices to 1985. Washington, D.C.: Gov't Print. Office, 1977. 1v.

844 IPT product price service. (In Pet. Times, vol. 83[2103], July 15, 1979, pp. 42-43)
Presents tabulated statistical information on prices.

845 IPT product price service. (In Pet. Times, vol. 83, Dec. 1979, pp. 30-31)
Gives statistical table of prices.

846 IPT production price service. (In Pet. Times, vol. 83, Oct. 1979, pp. 33-34)
Presents a tabulation of statistical information on prices.

847 The Iranian crisis: its impact on the price of oil. (In Energy Policy, vol. 7[2], June 1979, pp. 163-164)
Charts: Effect of oil price on consumption; Effect of oil price rise in 1979 on oil consumption; Effect of oil price rise in 1979 on oil's share of total energy consumption.

848 The Iranian petroleum supply distribution: an assessment in

terms of world oil prices, by M. Rodekohr and S. Sitzer. Washington, D.C.: Department of Energy, 1979. 51p.

849 Iraqi move adds to OPEC price turmoil. (In Oil & Gas J., vol. 77[25], June 18, 1979, p. 55)
Its crude oil customers told surcharges will match premiums.... Action highlights need for coordination of OPEC pricing structure.

850 Long-term oil prices forecast, by M.A. Adelman. (In Pet. Tech., vol. 21, Dec. 1969, pp. 1515-1520)
Forecast relates to the world oil market outside the US and covers the next 15 years. Forecast is in four steps: 1, Determine what the price would be under purely competitive supply and demand; 2, Project the price into the future; 3, Look at interference with supply and demand; 4, Predict the future of this influence.

851 The lookout for oil prices in the medium term, by G.F. Ray and C. Rowland. Surrey: University of Surrey, 1980. 15p.

852 Market prices. (In Pet. Times, vol. 83[2098] May 1, 1979, pp. 22-24)
Gives tables for United Kingdom, Austria, Belgium-Luxembourg, Denmark, Italy, France, Netherlands, Sweden, Switzerland, West Germany, Middle East, Far East and the Caribbean.

853 Monthly oil market review: spot crude oil prices. (In OPEC Bull., May 1980, pp. 20-43)
Contents: Petroleum product spot prices--1, The European market; 2, The United States and Caribbean markets; 3, The East of Suez market; 4, Freight market.# Tables: 1, OPEC spot crude oil prices, 1979-80; 2, Selected non-OPEC spot crude oil prices, 1979-80; 3, European market-product prices; 4, European market-basis Italy: product prices; 5, US market-product prices; Caribbean cargoes--FOB: product prices; 7, Gulf--FOB: product prices; 8, Singapore cargoes--product prices.

854 Monthly oil market review: spot crude oil prices. (In OPEC Bull., June 1980, pp. 47-81)
Includes nine graphs and ten tables.

855 Monthly oil market review: spot crude oil prices. (In OPEC Bull., July 1980, pp. 45-72)
Includes seven graphs and ten tables.

856 Monthly oil market review: spot crude oil prices. (In OPEC Bull., Aug. 1980, pp. 36-64)
There are eleven tables giving figures for the first half of 1980.

857 More surcharges hike OPEC prices. (In Oil & Gas J., vol. 77[22], May 28, 1978, pp. 56, 58)
Members, in midst of price surcharge spiral, hike long-term contract prices to more than $20/bbl.

858 New OPEC crude pricing methods suggested. (In Oil & Gas J., vol. 77[21], May 21, 1978, p. 25)
Contends that crude prices charged by the Organization of Petroleum Exporting Countries should be adjusted regularly in conjunction with an index based on the weighted dollar export prices of major OPEC trading partners. Table shows how real prices for OPEC crude have declined.

859 No relief in sight for crude supply, prices. (In Oil & Gas J., vol. 77[21], May 21, 1978, pp. 19-21)
Discusses Iranian oil price hike and price impact. Table: Breakdown of OPEC production capacity.

860 North Sea energy wealth, 1965-1985: oil and gas in the British and Norwegian economies, Vol. 1, by C. Johnson. London: Financial Times, 1978. xii, 163p. illus.
An international management report.# Partial Contents: 2, Oil and gas production and depletion--depletion and prices (present and future oil values and interest rates--risks of waiting--oil in the ground compared with other forms of investment--divergent interests of oil companies and governments); 3, Oil and gas prices and the value of North Sea output--a model of the world oil price market, World prices and North Sea production, Gas prices and North Sea, Allocating the benefits of low costs; 5, Offshore investment--investment in oil and gas, The pattern of capital expenditure; Cost escalation, Operating costs.

861 North Sea energy wealth, 1965-1985: oil and gas in the British and Norwegian economies, vol. 2, by C. Johnson. London: Financial Times, 1979. vi, 187p. illus.
An international management report.# Partial Contents: 7, Finance, banking and insurance--Types of North Sea lending, Capital expenditure and cash flow, Exploration finance, Offshore insurance; 8, The delicate balance of taxation--How governments take their share of oil revenues, The British system of oil and gas taxation, The Norwegian system of oil and gas taxation; 9, National accounts of the North Sea--North Sea accounts (United Kingdom shelf), The North Sea in the context of the UK economy, The North Sea and the Scottish economy, North Sea accounts (Norwegian Shelf); 10, How to spend it--The allocation of North Sea revenue, Allocation of North Sea taxes (United Kingdom), Allocation of North Sea taxes (Norway).

862 North Sea oil: the application of development theories. Brighton: University of Essex, Institute of Development Studies, 1976. [2], 50p. illus.

Conclusions include consideration of policy issues on pricing, and use of oil revenues.

863 Oil and the US dollar, by F.E. Banks. (In Energy Econ., vol. 2[3], July 1980, pp. 142-144)
Argues that the decline in the value of US currency has been a direct result of the drastic increase in the dollar value of US oil imports and the subsequent efforts to maintain US production and unemployment levels.# Headings: Oil prices and the dollar; US monetary expansion; Shortage of investment projects.

864 Oil market review (Jan.-Aug. 1980): stock development in five major oil consuming countries. (In OPEC Bull., Oct. 1980, pp. 36-76)
Countries considered are United States of America, Japan, France, United Kingdom and the Federal Republic of Germany.# Partial Contents: Spot crude oil prices; Petroleum product spot prices. Includes eleven graphs and fourteen tables.

865 Oil market review (Jan.-Sept. 1980): spot crude oil prices. (In OPEC Bull., Nov. 1980, pp. 35-69)
Partial Contents: Petroleum product spot prices; Freight market; The evolution of the gross product worth and the price differential of some selected OPEC crudes.... Gross product worth, the price differential. Includes eight graphs and twelve statistical tables.

866 Oil-price chaos squeezes many domestic refiners, by W.A. Bachman. (In Oil & Gas J., vol. 77[45], Nov. 5, 1979, pp. 29-33)
Tables: How crude cost after entitlement can vary; How refiners' costs imported crude has risen; Crude costs by class of refiners before and after entitlements.

867 Oil price hikes: as interpreted by the catastrophe theory, by W.P.S. Tan. Cheshire: Energy Price Consultants, 1980. [2], 12p.
Contents: Catastrophe theory of prices; OPEC and the oil industry; Influence of US and world economy; Dynamics of substitution; 1979/80 price hikes; 1979 substitute costs; The rise in the gold price.# Partial Charts: OPEC official reserves of foreign exchange; US consumer price index; Cumulative $ exchange loss since 1969; 1973/74 official price for Saudi market crude and auctioned price for Iranian light crude.

868 The oil price revolution of 1973-1974, by S. El Serafy. (In J. Energy Dev., vol. 4[2], Spring 1979, pp. 273+)

869 The oil price rise and the development of the world's fossil fuel resources: an address to the Australian Academy of

Technological Sciences Canberra, 18 Oct. 1977, by P.B. Baxendell. [Royal Dutch/Shell Group of Company]. 1977. 12p.

870 Oil: the airline predicament, by A. Thomson. (In ITA Bull., vol. 25[7], July 1980, pp. 569-573)
Headings: Fuel availability; Fuel price; Recovery of fuel price increases from the customer.

871 OPEC crude prices to take another leap, by R. Vielvoye. (In Oil & Gas J., vol. 77[27], July 2, 1979, pp. 60-61)
Table shows how OPEC crude prices have risen.

872 OPEC loses control of prices. (In Pet. Econ., vol. XLVII[1], Jan. 1980, pp. 2-3)
Shows table of crude oil: official sale prices--US dollars a barrel.

873 OPEC prices. (In Pet. Times, vol. 83, Dec. 15, 1979, p. 26)
OPEC prepares to raise crude oil prices for 1980's first half.# Tables: Saudi marker crude prices, 1970-1979; OPEC oil revenues, 1978.

874 OPEC prices look firm for balance of the year. (In Oil & Gas J., vol. 77[38], Sept. 17, 1977, pp. 31-33)
Table shows part of the oil supply/demand pricture in industrialized countries.

875 OPEC prices rise by instalments. (In Pet. Econ., vol. XLVI [1], Jan. 1979, pp. 2-3)
Headings: Spot market strength; Products prices; Counting the cost.

876 Passthrough of banked costs prodding US gasoline prices. (In Oil & Gas J., vol. 77[24], June 11, 1979, p. 30)
Considers ceiling price and banked costs.

877 Petroleum liquids in energy supply and demand: some significant influences, by M.J. Owings. (In J. Pet. Tech., vol. 24, May 1972, pp. 521-529)
Topics: Varying characteristics of energy markets; Natural gas price regulation; Coal economies of production.# Charts: Free world energy consumption, 1950-1985; US energy and gross national product, 1929 through 1968; Free world energy consumption, 1950-1985, by region; Total US energy consumption--percent by type of fuel; Per capita energy consumption vs. per capita income; US petroleum demand, 1950-1985, by use; US coal production, consumption and mine mouth price, 1956-1970.# Tables: 1, 1969 US average wellhead and mine mouth values; 2, US oil producing costs.

878 Price confustion and instability. (In Pet. Econ., vol. XLVII [2], Feb. 1980, pp. 46-47)
What is the typical oil refiner paying for a barrel of crude at the start of the 1980s? The answer is that he may be paying anything between $20 and $40 depending on the quality of the oil, the country of origin and whether he is buying on a contract or on the spot market. Includes a table of crude oil official sale prices.

879 Price hikes urged for Canadian domestic oil. (In Oil & Gas J., vol. 77[39], Sept. 24, 1979, pp. 86, 88)
Comment on a special report ordered by the Prime Minister, Joe Clark, which urges the Canadian Government to order immediate and rapid increases in the price of oil sold on the Canadian market.

880 The price of crude oil in the international energy market: a political analysis, by H. Maull. (In Energy Policy, vol. 5 [2], June 1977, pp. 142-157)
Topical Headings: Price strategies; Economic variables; Price changes since 1948--the main actors; Price reductions; Falls in posted prices and rise of OPEC.# Tables: 1, The evolution of international oil prices; 2, Distribution of resources in the international oil market; 3, Distribution of surplus profit: government and company take, 1948-1975 ($ per barrel).

881 The price of oil, by P. Desprairies. (In Revue de l'Energie, Aug./Sept. 1979, pp. 663+)

882 The price the market will bear. (In Pet. Econ., vol. XLVI [5], May 1979, pp. 189-183)
Headings: New price structure; Demand for OPEC supplies.

883 Realities of the oil business. (In Pet. Econ., vol. 46[8], Aug. 1979, pp. 310-311)
Table: Crude oil: official sales prices, dollars a barrel, 1978-1979.

884 Reflections on OPEC oil pricing policies, by H. Askari and M. Salehizadeh. (In OPEC Rev., vol. 3[1], Mar. 1979, pp. 21+)

885 Some long-term problems in OPEC oil pricing, by R.S. Pindyck. (In J. Energy Dev., vol. 4[2], Spring 1979, pp. 259+)

886 Tertiary oil needs front-end money and free-market prices: editorial. (In Oil & Gas J., vol. 77[37], Sept. 10, 1979, pp. 89+)

887 UK exports decline in wake of OPEC price increase: consum-

ing nations are forced to tighten their belts, by P. Algar. (In Pet. Rev., vol. 33[388], Apr. 1979, p. 18)
Discusses Britain's recent economic ills.

888 United States heating oil price debate. (In Pet. Econ., vol. 46[10], Oct. 1979, pp. 430-432)

889 US oil field inflation seen slowing. (In Oil & Gas J., vol. 77 [39], Sept. 4, 1979, pp. 82-83)
Includes a table showing how inflation erodes the price of US crude.

890 What are the sales prices of old refineries? by W. L. Nelson. (In Oil & Gas J., 77[37], Sept. 10, 1979, pp. 150+)
An answer to the much wider question--We note your use of replacement cost as a way to value a refinery. How does this compare with actual sale prices?# Table shows sale of 13 refineries, 1967-1978.

891 What price oil? Computer prediction of oil price rise. (In Energy Manager, vol. 2[2], Mar. 1979, p. 15)
With the loss of Iran's six million barrels a day production, the question is asked, How much will be paid per barrel of crude. With the aid of a computer, an attempt is made to find out.

892 World price of oil: a prospective view, by E.W. Erickson and H.S. Winokur. (In 2nd. Proc. AIChE., vol. 1, 1977, pp. 141-146)
Contends that the cost and benefit associated with any country's evaluation of alternative energy policies must commence with the world price of oil and the terms under which oil is available. Analysis forms the basis for evaluation of prospective potential price movements in the world oil market.

Statistics

893 The continuing world oil crisis: a statistical guide, by P. Algar. (In Pet. Rev., vol. 33[392], Aug. 1979, pp. 9-11, 13-15, 17-18)
Presents a statistical appraisal, listing 39 tables, each supported by minimal text.

894 Estimated costs of drilling and equipping wells. (In Offshore, vol. 39[7], June 20, 1979, pp. 67-68, 71)
Presents tabulated statistical tables.

895 IPT product price service. (In Pet. Times, vol. 83, June 1, 1979, pp. 31-32)
Gives statistical tabulation of prices.

896 Market data: more construction on petroleum. (In Can. Chem. Process, vol. 55, Sept. 1971, p. 56)
Presents statistics showing petroleum refinery, 1963-1970, and factory shipment, 1968-1970.

897 Market data: oil policy responds to attack. (In Can. Chem. Process, vol. 54, Sept. 1970, p. 62)
Presents statistics showing petroleum refining, 1963-1969; Materials used, 1966-1967; Imports, 1968-1969; Exports, 1968-1969.

898 Monthly oil market review: spot crude oil prices. (In OPEC Bull., Sept. 1980, pp. 36-62)
Gives statistical tables.

899 North Sea oil and gas statistics, by D. Duffy. (In Pet. Rev., vol. 33[391], July 1979, pp. 31+)
Six-monthly survey for Jan.-June 1979 North Sea activity with information supplied by the companies concerned.

900 Oil consumption rising to 1975. (In Pet. Eng., vol. 42, Jan. 1970, pp. PM2-3)
Charts: Future oil consumption by area; Capital outlays of Free World Petroleum Industry; Estimated world oil consumption, 1968-1975.

901 OPEC revenues to show big rise in 1979. (In Pet. Econ., vol. XLVI[6], 1979, p. 224)
Table: OPEC oil production, exports and revenues, 1974-1978.

902 Petroleum statistics. Luxembourg: Statistical Office of the European Communities, 1977. 64p.
Gives balance sheets for crude oil and petroleum products for 1976 and 1977 for ECC.

903 Profits half expansion needs: 28th annual report survey, by E. Adams. (In Pipeline & Gas J., vol. 201, June 1974, pp. EM2, 4-7, 10)
Tables: 1, Expenditure for facilities and reserves--US and non-communist countries; 2&3, Financial statistics of US petroleum companies; 4&5, Financial statistics of natural gas companies; 6, Drilling results by companies; 7, Natural gas and NGL products; 8, Liquid hydrocarbon producers among US companies.

904 Rapid growth for oil company earnings. (In Pet. Econ., vol. 46[12], Dec. 1979, pp. 505-506)
Tables: US oil companies; net income; earnings: major groups.

905 Refinery permit delays evaluated, by R. L. Sanson and D. Vio-

lette. (In Oil & Gas J., vol. 77[17], Apr. 23, 1979, pp. 78-83)

Partial Tables: Baseline costs for 200,000 b/d refinery; Product state and revenues; Potential capital and operating cost tradeoffs to avoid delays; Impact of delays with rising construction costs; Tradeoffs to avoid delays under differing inflation rates.

Taxation

906 Barber moves to check North Sea tax losses. (In Engineer, vol. 236, Mar. 8, 1973, p. 7)

The announcement by the Chancellor of the Exchequer in the House of Commons about crash government action to end the tax free bonanza on North Sea oil and gas.

907 Battle starts in Congress over excise tax. (In Oil & Gas J., vol. 77[20], May 14, 1979, p. 82)

Hearing of Congressional Committee of President Carter's proposed phased oil price decontrol with an excess profits tax.

908 Effects of the 1969 tax reform act on petroleum property values, by G. Dutton. (In J. Pet. Tech., Dec. 1970, pp. 1475-1479)

Topics: Percentage depletion; Statutory provisions; Taxable income on property unit; Depletable basis; Production payments; Statutory provisions; Development payments; Minimum tax.

909 IPAA arms to fight Carter excise tax. (In Oil & Gas J., vol. 77[20], May 14, 1979, pp. 84-85)

Reaction to President Carter's proposed excise tax: siphon off the bulk of the added income stemming from his oil decontrol plan.

910 North Sea oil taxes the sharing of risk: a comparative case study, by Ø. Bøhren and C. Schilbred. (In Energy Econ., vol. 2[3], July 1980, pp. 145-153)

The cash flow associated with North Sea oilfields are evaluated in terms of probability distributions of net present value. A simulation shows that despite structural differences the ability of the UK and Norwegian oil tax systems to distribute stochastic return between owners and government is remarkably similar.# Headings: Taxation in Norway and the UK; Evaluating uncertain multi-period cash flows; Changing taxes, financing and price uncertainty.

911 Offshore oil policy. London: Department of Energy, [1977]. 12p. (Energy Commission Paper, 19)

Concluding section of paper sets out issues in two areas of policy, taxation and development, and depletion.

912 Offshore petroleum resources: a review of UK policy, by S. Cochrane and J. Francis. (In Energy Policy, vol. 5[1], Mar. 1977, pp. 51-62)
Topical Headings: Fiscal measures--tax legislation, cost and tax structure; Energy pricing structure.# Partial Tables: 1, Oil costs and taxes in UKCS; UKCS balance of payments, 1976-80.

913 Oil from the UK continental shelf. London: Department of Energy, 1975. 5p. (Fact Sheet 1)
Partial Contents: Production costs; Government income--Royalty; Petroleum revenue tax (PRT); Corporation tax.

914 Petroleum revenue tax: statement ... to the House of Commons ... by J. Bennett. London: H.M. Treasury, 1978. 9 leaves.

915 Schlesinger defends administration's excise tax. (In Oil & Gas J., vol. 77[20], May 14, 1979, pp. 87-88)
Defense of the old decontrol plan.# Table: How US drilling, completion costs have risen.

916 Senate panel exempts heavy oil from excise tax. (In Oil & Gas J., vol. 77[38], Sept. 17, 1979, p. 46)
Comment on the proposed excise tax on decontrolled US crude.

917 Tax reform legislation: the mining and petroleum industry, by R.D. Brown. (In Can. Min. & Met. Bull., vol. 66, Jan. 1973, pp. 112-114; Mar. 1973, pp. 177-179)
Discusses Canada's new income tax legislation as it relates to the mining and petroleum industry. Includes table showing calculation of taxable income of mining and petroleum principal business corporations after 1976.

918 Taxing North Sea oil profits in the UK: special needs and the effects of petroleum revenue tax, by C. Rowland. (In Energy Econ., vol. 2[2], Apr. 1980, pp. 115-125)
The long lean time, high risk and excess profits that characterize offshore oil production set the North Sea aside from other UK industries. How these characteristics impair the incidence and effects of standard profits taxation achieved via Corporation Tax is shown.# Partial Tables: North Sea taxes as a percentage of total real resources generated in oil production from UK continental shelf; Three scenarios for North Sea oil prices at the UK coast; The field by field impact of PRT: after tax profitability; Discounted tax reliefs for a typical North Sea oilfield from royalties, PRT and Corporation Tax combined; Progressive North Sea oil taxation.

919 UK oil profits to attract bigger taxes. (In Noroil, vol. 7, Mar. 1980, pp. 24-25)

Includes a table showing North Sea profits, with projections of North Sea oil and gas contribution to leading oil companies in the UK sector. Quotes 1980 oil price assumption as circa $31.00 barrel.

VIII. SOLAR ENERGY

Costs

920 Analysis of systems for the generation of electricity from solar radiation, by W.G. Pollard. (In Solar Energy, vol. 23[5], 1979, pp. 279-392)
Partial Headings: Generating cost of solar electricity; Cost of storage and auxiliary power.# Capital cost of solar electric facilities is expressed in dollars for each kWh per yard of electrical output rather than dollars per kW of installed capacity as is customary for conventional electric generating plants.

921 Buying solar, by J. Dawson. Washington, D.C.: Gov't Print- Office, 1976. 1v.
A guide to the purchase of solar heating equipment in terms of available systems, including solar system costs.

922 Comparisons of deep well and insulated shallow earth storage of solar heat, Solar Energy, vol. 20 [2], 1978, pp. 127-137)
Four techniques for storing solar heat in earth are described by a hypothetical example. Cost comparison obtains optimal design parameters for each storage method and optimal surface area for an attached solar collector.

923 Conceptual development of a solar town in Iran, by M.N. Bahadori. (In Solar Energy, vol. 23[1], 1979, pp. 17-36)
Partial Tables: 4, The estimated initial costs of various solar energy convention methods for the solar town; 5, The estimated annual operating costs for various conversion methods of the solar town; 6, The cost of energy produced by various solar energy conversion methods.

924 Cost optimal development of mirrors associated with a high temperature solar energy system, by J.H. Hankins. (In Solar Energy, vol. 19[1], 1977, pp. 73-78)
Results are combined with a simple cost model to obtain a lower bound on the minimum cost per unit of redirected energy as a function of the unit mirror cost.

925 Cost reduction linked to using solar energy. (In Av. Wk., vol. 103, Aug. 18, 1975, p. 21)

Discusses a projection by the Energy Research and Development Administration's report that as much as 25 percent of the nation's electrical energy needs could be met by the year 2000 with solar energy if conversion equipment costs can be reduced sufficiently.

926 Current costs of solar powered organic rankine cycle engines, by R.E. Barber. (In Solar Energy, vol. 20[1], 1978, pp. 1-6)
Presents the technical and cost aspects of the organic rankine cycle and its interaction with the solar collector as a power system. When comparing the cost of solar power systems, it is shown that the dominant factor in the system cost is the collector cost. Consequently, low-cost collectors are crucial for commercialization of solar rankine systems. # Partial Tables: Estimated rankine cycle installed costs for production units; Solar power system estimated installed cost; Breakeven collector cost ratio.

927 Design and cost analysis for an ammonia-based solar thermochemical cavity absorber, by O.M. Williams. (In Solar Energy, vol. 24[3], 1980, pp. 255-233)
A design and cost analysis is introduced for a solar thermochemical cavity absorber operated at the focus of a tracking paraboloidal concentrator and based on the ammonia dissociation reaction.

928 Design considerations for residential solar heating and cooling systems utilizing evacuated tube solar collectors, by D.S. Ward and J.C. Ward. (In Solar Energy, vol. 22[2], 1979, pp. 113-118)
As solar heating systems become a commercial reality, greater efforts are now being employed to incorporate solar cooling components in order to obtain a complete solar heating and cooling system and thus take advantage of the cost effectiveness of year-round use of the solar equipment.

929 Development of low cost silicon crystal growth techniques for terrestial photovoltaic solar energy conversion, by J.A. Zoutendyk. (In Solar Energy, vol. 20[3], 1978, pp. 249-250)
Single crystal silicon solar cells are potential elements of large-scale solar energy conversion systems, but the current costs of these cells are too high. Paper reviews a US research program aimed at reducing the cost of silicon cells.

930 High efficiency low cost solar cell power, by I. Bekey and W. Blocker. (In Astronaut & Aeronaut., vol. 16, Nov. 1978, pp. 32-38)
Contends that a new concept for solar power modules could dramatically increase cell efficiency to lower the cost of spacecraft power, boost electric propulsion, and cut elec-

tric bills. Graph shows costs of different solar cell power modules.

931 Low-cost encapsulation materials for terrestial solar cell modules, by E.F. Cuddihy, B. Baun and P. Wills. (In Solar Energy, vol. 22[4], 1979, pp. 389-396)
Contends that solar cell modules must undergo dramatic reductions in cost in order to become economically attractive as practical devices for reducing electricity.# Partial Tables: Cost of substrates and superstrates based on calculations described in appendix 1; Low cost encapsulation materials for terrestial solar cell modules.

932 Low-cost solar array project. LSA field test--Annual report. Pasadena, Cal.: Jet Propulsion Laboratory, 1979. 50p. (DOE-JPL-1012-38)

933 Low-cost sun radiator for British market, by J. Summar. (In Engineer, vol. 240, May 8, 1975, p. 15)
Describes it as a simple, cheap solar collector for heating water using the sun's radiation.

934 New technologies for solar energy silicon: cost analysis of BCL process, by C.L. Yaws and others. (In Solar Energy, vol. 24[4], 1980, pp. 359-365)
New technologies for producing polysilian are being developed to provide lower cost material for solar cell which converts sunlight into electricity. Paper presents results for BCL Process (Battelle Columbus Laboratories); Cost sensitivity and profitability analysis results are presented.# Partial Tables: Cost analysis checklist for BCL process; Estimation of product cost for BCL process; Cost and profitability analysis summary for the BCL process.

935 New technologies for solar energy silicon: cost analysis of UCC silane process, by C.L. Yaws and others. (In Solar Energy, vol. 22[6], 1979, pp. 547-553)
Cost, sensitivity and profitability analysis are presented based on a preliminary process design plant.# Heading: Preliminary cost/sensibility analysis.# Tables: Cost analysis of UCC silane process; Estimation of product cost for UCC silane process; Cost and profitability analysis of UCC silane process.

936 The optimisation of solar timber drier using an adsorbent energy store, by N.A. Duffie and D.J. Close. (In Solar Energy, vol. 20[5], 1978, pp. 405-411)
Concerned with determining the optimized design of a solar timber drier equipped with an adsorbent energy store. For the cost data used, the adsorbent store gave lower total costs than did a gravel bed store.# Partial Heading: Annual system cost--initial investment; Changes in investment; Operating costs.# Tables: Cost of timber kiln; Cost

of solar energy system; Summary of optimization results/ energy costs.

937 Parametric cost analysis of photovoltaic systems, by V. Evtuhov. (In Solar Energy, vol. 22[5], 1979, pp. 427-433)
Cost analysis of photovoltaic systems based on different design philosophies was carried out. Analysis takes into account the fixed cost involved in photovoltaic installation--such as site preparation and inversion equipment. Confirms many of the accepted conclusions, presents them in a concise form and, in addition, allows one to see the relationship between the various system design regimes and sensitivity produced and the cost of the various system components.

938 Performance of low cost solar reflectors for transferring sunlight to a distant collector, by R.C. Zentner. (In Solar Energy, vol. 19[1], 1977, pp. 15-21)
Contends that the efficiency of reflection and optical transmission to a distant collector is a critical parameter, along with cost per unit area, in the selection of a heliostat design for the Central Collector Solar Electric Plant.

939 Perspectives on utility central station photovoltaic applications, by E.A. DeMeo and P.B. Bos. (In Solar Energy, vol. 21 [1], 1978, pp. 177-192)
Comparisons are made among a number of generic power plant conceptual designs, with the aid of an array design parameter that is defined to include array area related costs. Thinfilm approaches have potential for achieving these cost goals because of low material content and potentially low fabrication costs.

940 Sawing solar costs in half, by A. Weinstein and others. (In Solar Heating and Cooling, vol. 3[5], Oct. 1978, pp. 48-51)
Contends that there's no trick involved in cutting solar investment costs, and sets out to show how it can be done.

941 Simulation and cost optimization of solar heating of buildings in adverse solar regions, by B.D. Hunn and other. (In Solar Energy, vol. 19[1], 1977, pp. 33-34)
Presents a developed model which simulates the effects of hourly weather conditions on the performance of and cost of a combined solar/conventional heating system for buildings in cold, cloudy climates. The model exhibits the effects of several system and cost parameters on combined system cost so that optimal design can be determined.

942 A solar assisted heat pump system for heating and cooling residents, by B.W. Tleimat and E.D. Howe. (In Solar Energy, vol. 21[1], 1978, pp. 45-54)
An estimated cost of equipment and of its operation is compared with the cost of owning and operating fuel and

electrically heated systems. The conclusion is that the solar assisted heat-pump system with current fuel prices can provide immediate economic benefit over the all-electric home and is possibly on a par with residences using fuel oil or liquified petroleum gas, but it yields higher cost over systems using natural gas.

943 Solar coal gasification, by D.W. Gregg and others. (In Solar Energy, vol. 24[3], 1980, pp. 313-321)
Presents an evaluation of the technical and economic feasibility of solar coal gasification.# Partial Headings: Economics--Solar energy cost ($2.14/GJ or 2.26/$10^6$ BTU); Coal-oxygen energy cost ($3.22/GJ or 3.40/$10^6$ BTU); Product-gas energy cost. Table presents a comparative cost data for Lurgi and solar coal gasifiers.

944 Solar heat pump system: an analysis, by J.P. Hurley. (In Solar Heating & Cooling, vol. 3[3], June 1978, pp. 21-25)
Includes consideration that solar energy contributes to cost reduction.

945 Solar heating cuts bills, by 80 percent. (In Electronics & Power, vol. 21, Aug. 14, 1975, p. 809)
A solar heating bungalow designed by engineers at Loughborough University could cut central heating bills by up to 80 percent--a far greater amount than originally expected.

946 Solar satellites: space key to our power future, by G.R. Woodcock. (In Astron. & Aeronaut., vol. 15, July/Aug. 1977, pp. 30-43)
Presents issues and facts ... that solar power satellite is "too costly"--current estimates of total SPS costs equate to electricity costs at the ground output busbar in a range of $0.025-0.060 per kw/hr. This compares to a busbar cost of $0.030 per kw/hr for new oil-fired plants today. Includes a chart showing SPS cost uncertainties.

Economics

947 Econometric analysis of concentrators for solar cells, by A.S. Roy. (In Solar Energy, vol. 21[1], 1978, pp. 371-375)
Contends that when concentrating collectors are used with photovoltaic solar cells, the cost of the generated electricity is controlled by the concentrator cost. This enables the use of highly efficient cells and collectors of high concentration ratios, thereby reducing both the cell and the non-cell cost components per unit of generated electricity.

948 Economic analysis of solar, water and space heating, by US Energy Research and Development Administration. Washington, D.C.: Gov't Print. Office, 1976. 1v.
Cost comparisons are made for thirteen cities based on

local climatic conditions and cost alternatives using solar collectors costing $20, $15, or $10 per square foot.

949 Economics of solar energy, by E. L. Capener. (In Chem. Tech., vol. 6, Mar. 1976, pp. 190-193)
Presents data to compare a solar system for space and water heating with conventional gas and electrical heating systems. The basis for this comparison is a 15-year useful life for each alternative. A total cost of energy is calculated by amortizing the capital investment of each system over a 15-year life and adding to this the yearly maintenance and operating costs. Since there is a considerable variation in fuel costs and in the availability of solar energy, energy comparisons are done for several regions.# Topics: Today's cost of solar energy systems; Comparative system costs; Sensitivity of solar energy costs to economic incentives.# Charts: Effect of gas price increases on 15-year life heating costs for single-family residences.# Partial Tables: Solar energy system costs, heating only; Solar system cost breakdown; Installed cost for gas-fired furnaces, electrical furnaces, and water heaters; Consumer energy costs, 1974; Cost of energy for three heating options; Cost of energy for 50 percent solar/50 percent gas-heating systems with a 4 percent interest loan plus 22 percent reduction in system cost.

950 The economics of UK solar energy schemes, by P.F. Chapman. (In Energy Policy, vol. 5[4], Dec. 1977, pp. 334-340)
Discusses future fuel prices.# Tables: 1, Cost of supplying an extra unit of fuel at various times (1976 prices); 2, Net present value of fuel purchases (1976 prices) for 20-year period.

951 Load optimization in solar space heating systems, by C.D. Barley. (In Solar Energy, vol. 23[2], 1979, pp. 149-156)
Contends that when load variables, such as window and insulation types, are included in the economic optimization of a solar space heating system, the over-all cost is lower than that resulting from optimization of collection area for a fixed load.

952 A microeconomic approach to passive solar design: performance, cost, optimal sizing and comfort analysis, by S. Noll and W.O. Wray. (In Energy, vol. 4[4], Aug. 1979, pp. 575-591)
Presents a microeconomic methodology for the analysis and design evaluation of residential passive solar heating applications.# Partial Headings: Cost; Optimal sizing and financial performance--select results.

953 Payback of solar systems, by K.W. Boer. (In Solar Energy, vol. 20[3], 1978, pp. 225-232)

A variety of solar conversion systems is studied in a dynamic economical model in which the real cost of energy inflates. Payback times and dates of probable market entries are estimated.

954 Solar absorption cooling feasibility, by D. S. Ward. (In Solar Energy, vol. 22[3], 1979, pp. 259-268)
Contends that economic feasibility is heavily dependent upon the financial parameters assumed--in particular the inflation rate of conventional fuel costs. # Headings: Economic considerations; Economic analysis; Economic feasibility. # Table three presents economic assumptions which include: Capital cost of solar cooling subsystem; Capital cost of non-solar cooling system; Percentage of first year operating costs; Differential maintenance costs; First year cost of electricity; Inflation rate of operating costs; Inflation rate of maintenance cost; Investment tax credit; and, First year insurance rate on capital cost.

955 Solar heating systems for UK: design installation, and economic aspects, by S. J. Wozniak. London: HMSO, 1979, vii, 100p. [12]p. of plates.

956 Solar thermal electrical power plants for Iran, by S. Voidani and V. J. Woollam. (In Solar Energy, vol. 22[3], 1979, pp. 205-210)
Evaluates the production of electricity by solar thermal techniques for Iran. A simple economic analysis shows that due to the low domestic fuel cost of solar thermal power plants, most of the existing methods of producing electricity are more economical. Presents cost comparisons for power plants.

957 Technical and economic feasibility of ocean thermal energy conversion, by G. L. Dugger, E. J. Francis and W. H. Avery. (In Solar Energy, vol. 20[3], 1978, pp. 259-274)
Topical Headings: Conceptual designs for OTEC (Ocean Thermal Energy Conversion) plants; Technological status for OTEC plant components; Cost estimates for producing power. # Tables: Comparison of 238, 500 mt/yr aluminium plants and production cost changes over an 18-yr. period; Flow chart and estimated costs for the process from OTEC/ammonia through onshore, at-site decomposition and the electricity generated by fuel cells; Potential OTEC capacity and US fuel energy savings in the year 2000.

Finance

958 Determination of the optimal solar investment decision criterion, by M. R. Sedmak and E. M. Zampelli. (In Energy, vol. 4 [4], Aug. 1979, pp. 663-683)
Deals with the validity of the solar investment decision

criteria employed in various studies. Examines the life-cycle cost criterion commonly used by solar analysts.# Partial Tables: 2, Economic and cost assumptions; 3, Minimum annualized life-cycle cost solar system; 4, Maximum net present value system at time of present cost competitiveness.

959 The implementation of state solar incentives: financial programs, by J. Ashworth. Golden, Colo.: Solar Energy Research Institute, 1979. v, 57p.
Partial Contents: Review of the available financial incentives at the state level; State financial incentive initiatives in the period, 1973-1978; The experience of implementing financial incentives in selected states. Report examines the experience of selected states in implementing financial incentives for solar energy.

Management

960 Performance analysis and cost optimization of a solar-assisted heat pump system, by J.W. MacArthur, W.J. Palm and R.C. Lessman. (In Solar Energy, vol. 21[1], 1978, pp. 1-9)
Information collected was used to compute the payback time, based on cumulative costs, for each variation of the system's parameters when compared to a conventional system. The optimal combination of system components which had a payback time less than the mortgage life was determined. For the given initial costs of solar panels and storage reservoir, this optimal combination was found to be sensitive to the variations in mortgage and fuel cost growth rates.

961 Performance of combined solar-heat pump systems, by T. L. Freeman and others. (In Solar Energy, vol. 22[2], 1979, pp. 125-135)
A comparative study of the performance of combined solar heat pump systems for residential space and domestic hot water heating has been undertaken. Costs and the extent to which summer cooling is a requirement determine the relative merit of the conventional heat pump, conventional solar and parallel systems.

962 Solar electric generating system resources requirements, by R.C. Enger and H. Weichel. (In Solar Energy, vol. 23 [3], 1979, pp. 255-261)
The potential consumption of materials, land, water, manpower, energy, and money for four proposed electric generating system--a terrestrial solar thermal, a terrestrial photovoltaic, an orbiting solar reflector, and a satellite solar power system--is analyzed. Evaluation demonstrated that, per megawatt of electrical generating capacity, the terrestrial solar thermal system would require less

manpower, less energy of production, and less money than would the extra terrestrial system.

Pricing

963 Solar energy and US public utilities: the impact on rate structure and utilization, by C. Dickson, M. Eichen, and S. Feldman. (In Energy Policy, vol. 5[3], Sept. 1977, pp. 195-210)
Partial Topical Headings: Utility pricing and feasibility of solar energy systems--utility costs and pricing scheme efficiency.

Statistics

964 Statistical study of solar radiation information in an equatorial region--Singapore, by T.N. Goh. (In Solar Energy, vol. 22[2], 1979, pp. 105-111)
Summarizes results of regression analysis relating total solar radiation with common meteorological factors.

IX. WIND ENERGY

Costs

965 Analytical performance and economic evaluation of residential wind or wind and solar powered heating systems, by G. Darkazalli and J. G. McGowan. (In Solar Energy, vol. 21 [1], 1978, pp. 415-421)
Presents a performance and cost model for a variety of wind-powered space and water heating systems for single family residences.

966 Cost effectiveness of the vortex-augmented wind turbine, by O. Igra. (In Energy, vol. 4[1], Feb. 1979, pp. 119-130)
Presents cost estimates for conventional, horizontal-axis wind turbine.

967 Development of large wind turbine generators: a design feasibility and cost study, undertaken by a group consisting of British Aerospace Dynamic Group. London: Department of Energy, 1979. 20p. (WPG 79/3)

968 Federal wind energy program. Springfield, Va.: National Technical Information Service, 1975. 1v.
Includes consideration of wind energy system costs.

969 On the fluctuating power generation of large wind energy converters, with and without storage facilities, by B. Sørensen. (In Solar Energy, vol. 20[4], 1978, pp. 321-331)
Power fluctuations and time duration patterns of large hypothetical wind energy generators are analyzed using meteorological data for Denmark. A few remarks are made about the prospects for developing suitable storage facilities, and about the cost of total wind energy systems relative to other means of producing electricity.

Statistics

970 Statistical models for wind characteristics at potential wind energy conversion sites: final report. Evanston, Ill.: Northwestern Univ., Department of Civil Engineering, 1979. 86p. (DOE-ET-20283-1:EY-76-S-06-2342)

Presents simple models and guidelines for wind energy site surveys.

971 Wind energy statistics for large arrays of wind turbines. (New England and Central US regions), by C.G. Juslus. (In Solar Energy, vol. 20[5], 1978, pp. 379-386)

APPENDIX A

STATISTICAL REFERENCES

Note: This section contains publications which provide useful statistical information on energy.

Coal

972 Coal Age. New York: McGraw-Hill, 1911+
Gives coal price information and finance.

973 Investment in the Community coalmining and iron and steel industries: report on the 1980 survey. Brussels: European Coal and Steel Community Commission, 1980. 127p.
Gives statistical information. Shows that capital expenditure in the Community coal industry has risen from 932 million EUA (European Unit Account) in 1978 to 1167 million EUA in 1979.

974 Survey of Mines. Toronto: Financial Post. 1938+ Annual.
Gives financial and related statistical data on Canadian mines.

Electricity

975 All-electric homes in the United States, by US Federal Energy Information Administration. Washington, D.C.: Gov't Print. Office. 1964+ Annual.
Reviews the trends and levels of national and state bills, trends and levels of community bills.

976 Annual bulletin of electric energy statistics for Europe. New York: United Nations, Economic Commission for Europe. 1955+
Includes information on finance and international exchange of electric energy.

977 Annual statistical report, by Rural Electrification Administration. Washington, D.C.: Gov't Print. Office, 1952+

Gives financial and statistical information about the operations of REA electric borrowers.

978 The efficient use of electric motor drive systems, by D.E. Knight. Leatherhead, (Surrey): Electrical Research Association, 13p. illus.
Department of Energy statistics show that 86.1 TWh of electricity were supplied in the UK in 1976 at a cost of about £1.340 million. ERA estimates that, on average, around 65 percent of this total is consumed by the electric motor drives, indicating the importance of this one application of electricity in industry.

979 Gas turbine electric plant construction cost and annual production expenses. Washington, D.C.: Gov't Print Office, 1972+
Gives total number and dollar value of construction expenditure.

980 Handy Whitman index of public utility construction costs. St. Paul, Minn.: Whitman Requart and Associates. 1924+
Gives statistical information on cost trends for electric and gas utilities covering the United States in six geographical divisions.

981 Hydroelectric plant construction cost and annual production expenses. Washington, D.C.: Gov't Print. Office, 1957+

982 Pocket book of electric utility industry statistics. New York: Edison Electric Institute, 1955+
Gives statistical information on the electric utility industry, including capital and finance, sales and revenues, prices.

983 Preliminary figures, income, expenses, sales (Class A & B electric companies), by US Federal Energy Information Administration, Office of Accounting and Finance. Washington, D.C.: FPC Office of Public Information, 1976. 1v.
Contains financial and related data on privately owned utilities.

984 Steam-electric plant construction cost and annual production expenses. Washington, D.C.: Gov't Print. Office. 1974+ Annual.
Gives statistical information on investments and production costs, fuel use and costs....

Energy

985 Geothermal energy, by E.R. Berman. Parkridge, N.J.: Noyes Data Corp., 1975.

Contains a world survey of major geothermal installations, and cost factors.

986 Interfuel competition. US Congress. Senate. Judicial Committee. Subcommittee on Antitrust and Monopoly. 94th Cong. 1st Session. Washington, D.C.: Gov't Print. Office, 1976. 1v.
Gives statistical data on natural gas or petroleum producers and refiners who own interest in other energy industries.

987 Short-term energy forecasts, 1977-8. Department of Energy, Economics and Statistics Division, 1977. 3p. (Energy Commission Paper, no. 3)
Includes a table showing UK Gross Inland Primary Fuel Consumption, 1973/78.

Fuels

988 Annual summary of cost and quality of steam-electric plant fuels, with a supplement on the origin of the coal delivered to electric utilities. Washington, D.C.: Gov't Print. Office, 1976+ Annual.
Gives information on the cost and quality of fossil fuel delivered to steam electric generating plants.

989 Fuel cost and consumption: certificated route and supplemental air carriers, domestic and international operations. Washington, D.C.: Civil Aeronautics Board. 12 months ended Dec. 31, 1977 and 1976.

990 Handbook of airline statistics. Washington, D.C.: Gov't Print. Office. 1973+
Includes statistical information on certified air carriers --comparative traffic and financial data including consumption of various types of fuel.

991 A staff report of cost and quality of fuels for steam electric plants. Washington, D.C.: Gov't Print Office, 1975.
Gives information on the volume of coal and gas deliveries, and average prices.

Gas

992 Activities of major pipeline companies, by US Federal Power Commission. Office of Accounting and Finance. Washington, D.C.: FPC Office of Public Information. 1976+
A journal containing information on financial and statistical data on interstate natural gas pipeline companies.

993 Annual report for natural gas pipeline companies (Class ABC & D). Washington, D.C.: FPC Office of Public Information. 1976+
Gives statistical and financial data on natural gas pipeline companies.

994 Average prices received and paid by major pipelines for gas. Washington, D.C.: FPC Office of Public Information. 1976+
Irregular. Information given is about prices and revenue data.

995 Forecast of capital requirements of US gas utility industry. Arlington, Va.: 1978+

996 Main line natural gas sales to industrial users. Washington, D.C.: Department of the Interior. 1974+ Annual.
Gives statistical information on natural gas main line (direct) transmission sales to individual users, showing price for individual companies.

Oil

997 Investment research: offshore petroleum exploration notes. London: Stock Exchange, n.d.
Irregular. For private circulation only to members of the Stock Exchange. Covers operations in territorial waters, giving stock symbol, share prices, shares issued and market value.

998 Oil energy statistics bulletin. Babson Park, Mass.: Oil Statistics Co., 1923+ Biweekly.
Gives statistical analysis of world-wide energy developments and financial and stock market data.

999 OPEC special fund. Vienna: OPEC, 1976+ Annual.
Gives details of OPEC aid to developing countries.

1000 Petroleum industry, by US Congress. Judiciary Committee. Subcommittee on Antitrust and Monopoly. 94th Cong. 1st Session. Washington, D.C.: Gov't Print. Office. 1975.
Gives information on financial data and rates of return, and share ownership of the six largest US oil companies.

1001 Platt's oilgram price service. New York: McGraw-Hill, 1923+ Daily.
Presents tables of prices for various petroleum products around the world, as well as spot tanker rates and market conditions.

1002 Reserves of crude oil, natural gas liquids and natural gas

in the United States and Canada as of ... Washington, D.C.: American Petroleum Institute. Annual.
Gives statistical information.

1003 Survey of oil. Toronto: Financial Post. 1929+ Annual.
Gives financial data on petroleum companies listed on the Canadian stock exchange.

1004 World offshore and gas: a review of offshore activity and an assessment of worldwide market prospects for offshore exploration/production equipment and materials. Aberdeen: Scottish Council (Development of Industry), 1975. vi, 212p. illus. (maps)
Gives statistical and directory information.

1005 World wide crude oil prices. Washington, D.C.: Gov't Print Office, 1976+ Annual.
Gives information on foreign crude oil prices, including data on total posted price, landed cost charter, delivery price, average producing cost....

1006 The world petroleum market, by M.A. Adelman. Baltimore: Johns Hopkins Press, 1972.
Gives information on past behavior of oil prices, cost structures, production and transportation costs which mold price behavior, tanker rates, price movements, and price rise chronology.

APPENDIX B

Presented are the abbreviated forms of journal titles cited in the bibliography together with their full titles. Shown also are commencement date of publication, frequency, and place of publication.

Air Cargo Mag.
Air Cargo Magazine [1942+ (M)] Oak Brook, IL.

Air Pollution Control Assoc. J.
Air Pollution Control Association Journal [1951+ (M)] Pittsburgh, PA.

Air Transp. World
Air Transport World [1964+ (M)] Cleveland, OH.

Airc. Eng.
Aircraft Engineering [1929+ (M)] London.

Am. Chem Soc. Div. Fuel Chem. Papr.
American Chemical Society. Division Fuel Chemical Paper [1879+ (M)] Washington, DC.

Am. Dyestuff Reporter
American Dyestuff Reporter [1917+ (M)] New York.

Am. Gas Assoc. M.
American Gas Association Monthly [1919+ (M)] Arlington, VA.

Am. Soc. C.E. Proc.
American Society of Civil Engineers Proceedings [1956+ (M)] New York.

Am. Soc. for the Adv. of Science
American Society for the Advancement of Science [1880+ (W)] Washington, DC.

Ann. Rev. Energy
Annual Review of Energy [1976+ (A)] Palo Alto, CA.

Applied Energy
[1975+ (Bi-m)] Essex, England.

ASHRAE
American Society of Heating, Refrigerating and Air Conditioning Engineers. [1894+ (M)] New York.

Astronaut & Aeronaut
Astronautics and Aeronautics [1957+ (M)] New York.

Atom
[1964+ (10/yr)] Los Alamos.

Aust. Inst. Min. & Metal.
Australian Institute of Mining & Metallurgy Symposia Series [1972+ (Irreg)] Parkville, VIC.

Automotive Eng.
Automotive Engineer [1962+ (M)] Suffolk.

Aviation Week
Aviation Week and Space Technology [1916+ (W)] New York.

Brit. Airways Air Safety Rev.
British Airways Air Safety Review [n. d. (Irreg)] London

Building Systems Design <u>see</u> Energy Eng.

Bull. Atomic Scientists
Bulletin of the Atomic Scientists [1945+ (Bi-m)] Chicago, IL.

Business Comm. Av.
Business and Commercial Aviation [1958+ (M)] New York.

Can. Min. Met. Bull.,
Canadian Mining and Metallurgical Bulletin [1969+] Montreal.

Canadian Chem. Process.
Canadian Chemical Processing [1917+ (M)] Ontario.

Chem. & Eng. N.
Chemical and Engineering News [1923+ (W)] Washington, DC.

Chem & Ind.
Chemistry and Industry [1881+ (S-M)] London.

Chem. Eng.
Chemical Engineer [1923+ (M)] Rugby, England.

Chem. Eng. Prog.
Chemical Engineering Progress [1947+ (M)] New York, NY.

Chem. Process.
Chemical Processing [1938+ (14/yr)] Chicago, IL.

Chem. Tech.
Chemical Technology [1970+ (M)] Washington, DC.

Coal Age
[1911+ (M)] New York, NY.

Coal Min. Process.
Coal & Mining Processing [1964+ (M)] Chicago, IL.

Combustion
[1929+ (M)] New York, NY.

Cryogenics
[1960+ (M)] Guildford, England.

DE/J.
[1881+ (M)] Elmhurst, IL.

Diesel Equipment Superintendent
[1923+ (M)] Norwalk, CT.

Elec. Constr. Maintenance
Electrical Construction and Maintenance [1901+ (M)] New York, NY.

Elec. World
Electrical World [1874+ (S-M)] New York, NY.

Electrical Review
[1872+ (W)] London.

Electronics and Power
[1955+ (22/yr)] London.

Energy and Building
[1977+ (Q)] Switzerland.

Energy Comm.
Energy Communications [1975+ (6/yr)] New York, NY.

Energy Dev.
Energy Development in Japan [1978+ (Q)] Chicago, IL.

Energy Digest
[1952+ (Bi-m)] Watford, England.

Energy Econ.
Energy Economics [1979+ (Q)] Guildford, England.

Energy Eng.
Energy Engineering [1904+ (Bi-m)] Brooklyn, NY.

Energy Int.
Energy International [1963+ (M)] San Francisco, CA.

Energy Manager
[1978+ (M)] Guildford, England.

Energy Policy
[1973+ (Q)] Guildford, England.

Energy Systems & Policy
[1973+ (Q)] New York, NY.

Energy, The Int. J.
Energy, the International Journal [1976+ (Q)] Oxford, England.

Energy World
[1973 (M)] London.

Engineer
[1856+ (W)] London.

Engineering
[1866+ (M)] London.

Engineering & Mining J.
Engineering and Mining Journal [1975+ (M)] New York, NY.

Engineering J.
Engineering Journal [1918+ (4/yr)] Montreal, Canada.

Flight Int.
Flight International [1909+ (W)] London.

Gas World
[1884+ (M)] London.

Glass Technology [1960+ (Bi-m)] Sheffield, England.

Heating/Piping/Air Conditioning
[1929+ (M)] Cleveland, OH.

Heating Vent. Eng.
Heating and Ventilating Engineer [n.d. (10/yr)] London.

Hydrocarbon Process.
Hydrocarbon Processing [1922+ (M)] Houston, TX.

ICAO Bull.
International Civil Aviation Organization Bulletin [1946+ (M)] Montreal.

IEEE Proceedings
[1913+ (M)] Piscataway, NJ.

IEEE Spectrum
[1964+ (M)] Piscataway, NJ.

IEEE Trans. Ind. Appl.
IEEE Transactions Industry Applications [1965+ (Bi-m)] Piscataway, NJ.

IEEE Trans. Power Apparatus Syst.
IEEE Transactions Power Apparatus Systems [1884+ (Bi-m)] Piscataway, NJ.

Industrial Finishing
[1924+ (M)] Wheaton, IL.

Int. Atomic Agency Bull.
International Atomic Agency Bulletin [1959+ (6/yr)] Vienna, Austria.

Int. J. Energy Res.
International Journal of Energy Research [1977+ (Q)] Chichester, England.

Int. Pet. Times
International Petroleum Times [1899+ (Fortn)] London.

Interavia
Interavia: World Review of Aviation-Astronautics, Avionics [1946+ (M)] Cointin-Geneva.

Iron Age
[1855+ (W)] Radnor, PA.

ITA Bull.
ITA Bulletin (Institut du Transport Aérien) [1949+ (W)] Paris.

J. Energy Dev.
Journal of Energy Development [1975+ (S-A)] Boulder, CO.

J. Inst. Fuel
Journal of the Institute of Fuel [1970+ (A)] Belfast, Northern Ireland.

J. Inst. Nuclear Eng.
Journal of the Institution of Nuclear Engineers [n.d.] London.

J. Pet. Tech.
Journal of Petroleum Technology [1949+ (M)] Dallas, TX.

Lighting Design and Application
[1906+ (M)] New York, NY

Machine Design
[1929+ (28/yr)] Cleveland, OH.

Mechanical Eng.
Mechanical Engineering [1906+ (M)] New York, NY.

Mining Cong. J.
Mining Congress Journal [1915+ (M)] Washington, DC.

Mining Eng.
Mining Engineer [1949+ (M)] London.

Mod. Materials Handling
Modern Materials Handling [1946+ (M)] Denver, CO.

Noroil
[1973+ (M)] Stavanger, Norway.

Nuclear Eng. Int.
Nuclear Engineering International [1956+ (M)] Sutton, England.

Nuclear Engineer see J. Inst. Nuclear Eng.

Nuclear Engineering see Nuclear Eng. Int.

Nuclear Safety
[1959+ (Bi-m)] Oak Ridge, TN.

Ocean Industry
[1966+ (M)] Houston, TX.

Offshore
[1951+ (14/yr)] Tulsa, OK.

Offshore Eng.
Offshore Engineer [1975+ (M)] London.

Oil & Gas J.
Oil and Gas Journal [1902+ (W)] Tulsa, OK.

Oil & Gas J. Tech. see Oil & Gas J.

OPEC Rev.
OPEC Review [(Q)] Oxford, England.

Pet. Econ.
Petroleum Economist [1934+ (M)] London.

Pet. Eng. Int.
Petroleum Engineering International [1929+ (15/yr)] Dallas, TX.

Pet. Rev.
Petroleum Review [1947+ (M)] London.

Pipeline Gas J.
Pipeline and Gas Journal [1859+ (14/yr)] Dallas, TX.

Plant Eng.
Plant Engineering [1947+ (Fortn)] Barrington, IL.

Power [1971+ (7/yr)] Dana Point, CA.

Power Eng.
Power Engineering [1896+ (M)] Barrington, IL.

Proc. AIChE J.
American Institute Chemical Engineers Journal [1955+ (Bi-m)] New York, NY.

Process Eng.
Process Engineering [1972+ (M)] London.

Revue de l'Energie
[1949+ (M)] Paris.

Rock Prod.
Rock Products [1902+ (M)] Chicago, IL.

Sc. Am.
Scientific American [1845+ (M)] New York, NY.

Solar Energy
[1957+ (M)] Elmsford, NY.

Tappi
Technical Association of Pulp & Paper Industry [1949+ (M)] Atlanta, GA.

Textile Ind.,
Textile Industries [1899+ (M)] Atlanta, GA.

Turbomach. Int.
Turbomachinery International [1958+ (Bi-m)] Norwalk, CT.

Wireless World
[1911+ (M)] London.

World Oil
[1916+ (14/yr)] Houston, TX.

NAME INDEX

References are to entry numbers in the bibliography

SUBJECT INDEX

References are to entry numbers in the bibliography